To my good friend Doug.

Steve Connor

Science or Philosophy?

A Perspective on the Evolution of Life

Steven J. Connor

The Hamburg Press
North Augusta,
South Carolina

2001

Manufactured in the United States of America

Published in the United States of America by

The Hamburg Press
Post Office Box 984
Augusta Ga. 30903-0984

info@thehamburgpress.com
www.thehamburgpress.com
www.authorswebpress.com

Cover design & layout by Michele Blackburn

Library of Congress Cataloging-in-Publication Data

Connor, Steven J.,
Science or Philosophy? A Perspective on the Evolution of Life

SUMMARY: Clear language discussion of the lack of proof for the Theory of Evolution
[I. Evolution - Non-fiction]

ISBN 0-916587-04-5

Library of Congress Control Number: 2001090696

1. From Paley to Dawkins

The Evolution of Evolution

Relates the history of intellectual thought that led to naturalism and, more specifically, the biological theory of evolution.

2. Charles Darwin

His Theory and Its "Proofs"

Examines the theory of evolution in more detail, including its "proofs" as presented by evolution's protagonists.

3. Origin of Life

Investigates theories related to the origin of life from the primeval chemical soup.

4. Paley and Dawkins Again

The Argument for Intelligent Design

Presents reasons for believing that a Creator exists who made us and all we see.

5. The Attempt to Make Evolution a Fact

Examines hoaxes of the past that have attempted to finally provide the missing evidence for evolution. Also, presents some of the intimidating methods scientists and other intellectuals use to squash opposition.

6. Did Lyell Get It Right?

Issues in Geology

Investigates geological issues such as radioactive dating and catastrophism.

7. Richard Owen's Fossils

Special Challenges for Creationists

Raises issues that could be troubling to those who do not accept naturalism, such as human fossils and dinosaurs.

8. How Do We Know The Truth?

Is there an epistemology that works better than science?

9. References

Preface

The dominant worldview of science (and of many other intellectual pursuits) is that the universe, and all the life within it, created itself. This philosophy, known as naturalism (nature is all there is), has as its basis the desire of humans to explain their origins in terms of physical laws they can understand, without reliance on a superior being. There is no fundamental discovery in science that leads to naturalism; it is an *a priori* assumption. If nature is all there is, then the intellectual methodology known as science is the only key to understanding all of reality.

Given this status, science, in effect, has become a religion, and its priests are the scientists. Most religions have their mythology and supernatural aspects that form the basis for interpreting the world. In the case of naturalism, the big-bang cosmology, the chemical origin of life, and the Darwinian evolution of life are the founding principles and mythology of naturalism. None of these things have been observed to occur, but they are believed as strongly as the foundational tenants of any religion.

In the naturalistic world, the collective judgement of scientists is the final arbiter of what is truth. The scientists determine what questions will be asked, what research will be funded, and how experimental results will be interpreted. Any investigation into nature that does not assume the reality of naturalism is deemed to be as far outside science as is astrology. Religion is allowed in the world of naturalism, but it is declared to be a product of mankind's fertile mind and must be maintained separate from the only absolute approach to truth, science. Religion's sphere is limited to morality and ethics, topics that, although essential to a civilized society, contain no absolute truth. Religion and morality are personal preferences.

Scientists tell us that the coercive evidence for biological evolution makes any consideration of origins not based on naturalism pure folly. However, if naturalism guides the pursuit of science, then scientists cannot help but find evidence to support the presupposition of naturalism. But, what if the presumption is not true? What if much we have been taught in school and public television is not nearly so solid as the priests of naturalism would have us believe?

This socio-scientific condition is not suggestive of a conspiracy. Scientists are almost always honest men, seeking the truth of nature. But as humans, they are susceptible to the same errors in judgement and prejudices as all the rest of us. If their fundamental precept or belief is wrong, then they will quite honestly be led to believe a falsehood and vigorously teach others to do so.

There are many scientists, engineers, and educators who believe in the products of naturalism (such as biological evolution), because they were taught that way in school, and society reinforces that belief by labeling as ignorant anyone who does not so believe. Nevertheless, these majority of professionals have an ambiguity about them, because so many also believe in God to one extent or another. These scientifically educated individuals are not especially committed to evolution. Their jobs and their interests would not be greatly affected if evolution were to suddenly be declared false.

It is for these "*uncommitted*" intellectuals that this book is written. The book explores the concept of evolution and the origin of life to determine if the evidence is indeed coercive. It then moves on to the evidence for intelligent design in nature. The discussion is not hard-core science and makes no attempt to be academically rigorous. It is an argument of logic within the context of science. Neither does the discussion make much use of theological arguments (although there some very few Bible verses interspersed to enliven the discussion).

The author is indebted to the brillant works of Michael Behe, Michael Denton, and Phillip Johnson who are frequently cited throughout this book. Readers seeking more in-depth knowledge are encouraged to consult these authors. The general contents of the chapters are as follows:

From Paley to Dawkins - the Evolution of Evolution: Relates the history of intellectual thought that led to naturalism and, more specifically, the biological theory of evolution.

Charles Darwin - His Theory and Its "Proofs": Examines the theory of evolution in more detail, including its "proofs" as presented by evolution's protagonists.

Origin of Life: Investigates theories related to the origin of life from the primeval chemical soup.

Paley and Dawkins Again - The Argument for Intelligent Design: Presents reasons for believing that a Creator exists who made us and all we see.

The Attempt to Make Evolution a Fact: Examines hoaxes of the past that have attempted to finally provide the missing evidence for evolution. Many of these hoaxes are still knowingly maintained in textbooks today. The chapter also presents some of the intimidating methods scientists and other intellectuals use to squash opposition.

Was Lyell Right? - Issues in Geology: Investigates geological issues such as radioactive dating and catastrophism.

Richard Owen's Fossils - Special Challenges for Creationists: Raises issues, such as human fossils and dinosaurs, that could be troubling to those who do not accept naturalism.

How Do We Know Truth?: What is science and what are its methods? Is there an epistemology that works better than science?

One

From Paley to Dawkins

The Evolution of Evolution

The English scholar, William Paley, was born in 1743 to intelligent but frugal parents with ample opportunity to give their son an outstanding education. Paley became well known as an educator, philosopher, and theologian. Although not known as an originator of new ideas, he was apparently very successful as an author of text books. His most famous book, *Natural History - or, Evidences of the Existence and Attributes of the Deity, Collected from the Appearances of Nature,* was first published in 1802. In this book, he expounded with compelling logic the philosophy of Natural Theology, in which the wisdom, power, and goodness of

God were believed to be revealed by nature as taught in the Scriptures:

> *The heavens declare the glory of God; and the firmament sheweth His handiwork. Day unto day uttereth speech, and night unto night sheweth knowledge.* Psalms 19:1, 2

> *For the invisible things of Him from the creation of the world are clearly seen, being understood by the things that are made, even His eternal power and Godhead; so that they are without excuse.* Romans 1:20

In this book, Paley makes the argument that because no one assumes a complicated, well designed watch is a natural object, without a maker, neither is the complex, well designed world without a maker. Paley states (Dawkins 1996):

> In crossing a heath, suppose I pitched my foot against a stone, and were asked how the stone came to be there; I might possibly answer, that, for anything I knew to the contrary, it had lain there for ever: nor would it perhaps be very easy to show the absurdity of this answer. But suppose I had found a watch upon the ground, and it should be inquired how the watch happened to be in that place; I should hardly think of the answer which I had before given, that for anything I knew, the watch might have always been there.

> that the watch must have had a maker: that there must have existed, at some

> time, and at some place or other, an artificer or artificers, who formed it for the purpose which we find it actually to answer; who comprehended its construction, and designed its use.

> every indication of contrivance, every manifestation of design, which existed in the watch, exists in the works of nature; with a difference, on the side of nature, of being greater or more, and that in a degree which exceeds all computation.

Paley was not alone in his belief that complex, interacting systems demand the existence of an intelligent Designer. Most of the scientists of his day and before believed in a Creator. Their study of nature established for them, without doubt, that "*by the Word of the Lord were the heavens made; and all the host of them by the breath of his mouth...for He spake, and it was done; He commanded, and it stood fast.*" Psalms 33: 6, 9.

Although the concept of intelligent design seems eminently logical, today we find that very few scientists, and academics in general, believe in intelligent design. The most popular book on biological evolution since Charles Darwin's *Origin of Species* (1859), is Richard Dawkins', *The Blind Watchmaker, Why the Evidence of Evolution Reveals a Universe without Design* (1996). This book, treasured by scientists and laymen alike, directly attacks the concept of intelligent design - even within its title. Dawkins writes:

> Biology is the study of complicated things that give the appearance of having been designed for a purpose. (emphasis supplied)

Scientists That Believed in a Creator God

Not all those identified below were biblical literalists; however, all are thought to be theists who believed that a personal God created the universe. Many did indeed believe in a literal understanding of the Bible, including a six-day creation. None felt that belief in a Creator limited the pursuit of science.

Francis Bacon	1561–1626	Scientific method
Johannes Kepler	1571–1630	Laws of planetary motion
Blaise Pascal	1623–1662	Hydrostatics; laws of probability
Robert Boyle	1627–1691	Chemistry; gas dynamics
Isaac Newton	1643–1727	Calculus; laws of motion; optics; gravity
Carolus Linnaeus	1707–1778	Modern taxonomical system
William Herschel	1738–1822	Astronomy; discovered Uranus
Georges Cuvier	1769–1832	Comparative anatomy; paleontology
Michael Faraday	1791–1867	Electric field theory; electric generator
Charles Babbage	1791–1871	Conceived the first digital computer
Samuel Morse	1791–1872	Inventor of the telegraph
Louis Agassiz	1807–1873	Glacial geology; ichtyology
James Joule	1818–1889	Thermodynamics
George Stokes	1819–1903	Fluid mechanics
Gregor Mendel	1822–1884	Genetics
Louis Pasteur	1822–1895	Bacteriology; vaccination; pasteurization; immunization
Lord Kelvin	1824–1907	Thermodynamics; transatlantic cable
Bernhard Riemann	1826–1866	Non-Euclidian geometry
Joseph Lister	1827–1912	Antiseptic surgery
James Clerk Maxwell	1831–1879	Equations of electromagnetism
Lord Rayleigh	1842–1919	Acoustics; optics; wave propagation in fluids
J. J. Thomson	1856–1940	Discovery of the electron

Dawkins admits that the world, especially life, looks as if it were designed; Paley's logic is powerful and drives the uninitiated observer to conclude intelligent design. However, according to Dawkins, design is an illusion. The watchmaker analogy does not hold for living systems. In spite of all our experiences, living things can emerge from the non-living, and complex systems can develop naturally from simple things. Dawkins argues that those instructed in the power of natural selection, the driving force of biological evolution, can clearly see that all that exists in the world is the result of the mechanical operation of the laws of physics on matter. Dawkins states:

> All appearances to the contrary, the only watchmaker in nature is the blind forces of physics, albeit deployed in a very special way. A true watchmaker has foresight: he designs his cogs and springs, and plans their interconnections, with a future purpose in mind's eye. Natural selection, the blind, unconscious, automatic process which Darwin discovered, and which we now know is the explanation for the existence and apparently purposeful form of all life, has no purpose in mind. It has no mind and no mind's eye. It does not plan for the future. It has no vision, no foresight, no sight at all. If it can be said to play the role of watchmaker in nature, it is the blind watchmaker.

When Darwin published his famous book, most scientists were creationists and did not feel that creationism limited their pursuit of scientific knowledge. In the years following *Origin*, Darwin's theory was vigorously debated - not accepted automatically. Darwin and a few others such as Thomas Huxley had to sell the concept. Although, as we shall see in subsequent chapters, the theory of evolution was as scientifically weak when it was proposed as it is today, there was something very powerful in its favor. Evolution formed the scientific basis for the growing philosophy of naturalism (nature is all there is) and humanism (confidence in human virtue). Academics, and eventually others, began to see the value of a naturalistic viewpoint: without God, there is no accountability or limitation on our actions. Dawkins himself reveals the motivating power behind the theory when he says, "Darwin made it possible to be an intellectually fulfilled atheist."

How could Darwin turn the tide and overcome the self-evident argument from design? Why did not Dawkins draw the obvious conclusion - that something that appears to be designed may, in fact, be designed? A partial explanation is revealed by a study of the history of intellectual thought. Arguably the story begins with the Greeks, but we will begin the telling of it with the Middle Ages, progressing through the Renaissance, and into the modern age by way of the Enlightenment.

A Short Summary of Intellectual History

Intellectual Thought in the Medieval Ages

The Middle Ages get their name because historians of the Renaissance period noted that they were in an age of rebirth of the cultural glory of the Roman Empire. The roughly 1,000 years between the fall of Rome in the 400s and the awakening of the 1400s was seen as a dark, middle time. By the 400s, the peace enforced by the Romans had ceased to function in western Europe and would soon cease in the eastern part of the former

Roman Empire. Consequently, the cities, which were the centers of commerce, culture, and academia, being dependent on public order, became depopulated.

The only institution that remained during the Middle Ages was the Christian church. However, it was a church that had become partially infested with paganism by the likes of Emperor Constantine. The church was dedicated to keeping the people in ignorance so that ecclesiastical power could be maintained. What little intellectual life remained was within the hierarchy of the church and thoroughly controlled by it.

The intellectual climate of the Middle Ages was strongly religious, but more inclined to recognize the authority of the church than the authority of divine revelation in the Bible. The church strongly endorsed Greek science, a condition that continued to exist into the Enlightenment, as evidenced by Galileo's difficulties with the church. Superstition was rampant, given the half-pagan church's dominance of thinking. In science, there was little inquiry into how the natural world operated; reasoning from cause to effect was rare. Events in the natural world, which even the unlearned of today would recognize as the working of natural law, would be attributed to direct divine intervention. It is hard to imagine a situation more suitably designed to ultimately bring reproach on religion as subsequent empirical studies swept away the superstition to reveal a world largely dominated by natural law that people could mathematically describe.

There was no distinct demarcation between the Middle Ages and the Renaissance, especially since the later 300 years of the Middle Ages were characterized by increasing intellectualism as Thomas Aquinas and others rediscovered the full depth of Aristotle and Plato from their writings and applied subjects such as logic, mathematics, law, medicine, and astronomy to the issues of life.

Intellectual Thought During the Renaissance

The short transition from the Middle Ages into the Enlightenment, known as the Renaissance or rebirth, began in Italy, but it was carried into the rest of Europe through commerce and scholars who had studied in Italy. As with other periods of intellectual history, the Renaissance was manifest in a relatively small subset of the population: the intellectuals, the wealthy, and the explorers. However, their influence on the age was very great. The period was marked by a revival of the ancient classic humanism, proclaiming the dignity and self-reliance of mankind and the beauty of this earthly life. This attitude is strongly contrasted with the medieval emphasis on the depravity of man and the miseries of this world in comparison to the life to come.

The Renaissance man expressed a new individualism because of his rejection of the rigid social and religious structures that had enslaved mankind so long. This emancipation resulted in a rationalism and in a fresh attitude toward the study of nature that helped fuel the scientific revolution. Although the humanistic way of thinking that developed was antithetical to the Protestant Reformation, the individualism and moderate rationalism, the demanding of a rational reason for beliefs, and the scholarship required for detailed Bible study, must certainly have contributed to the Reformation (as did the Renaissance invention of the printing press), which followed on the heels of the Renaissance period.

Nicholas Copernicus was a Renaissance thinker of great importance. Prior to Copernicus' time, the earth was assumed to be the center of the universe. The heavenly bodies circled the earth in accordance with divine law, as plainly seen by casual observation of the sky. However, more careful observations by astrologers in the employ of the church and kings revealed that some of the bodies progressed at different rates, and at times moved backward when compared to the rest of the stars. These special case bodies were the planets, moons, asteroids, comets, and other non-stellar bodies. The astrologers invented numerous subtheories to explain planetary motion. Over time, the theories became so complex that the movement of the heavenly bodies became nearly incomprehensible. By the thirteenth century, the laws of heavenly motion were so complex that Alfonso X declared that if God had consulted him on the creation of the universe He would have received good advice.

In the mid 1500s, Copernicus proposed that the Sun was the center the of universe and that the planets moved about the sun in circular orbits. His stated purpose was to create a hypothetical system to simplify astrological predictions, not to overturn accepted church/Greek theories. Historians suspect that the hypothetical argument was a cover story for proposing a bold new theory in contradiction to established theology, but they are not certain.

Galileo Galilei, an Italian astronomer, announced in the early 1600s that Copernicus' hypothesis was not just a simplifying construct, but a true description of heavenly motion. Galileo supported his conclusion with observations and calculations. Unfortunately, Galileo also assumed circular orbits, and his calculations yielded somewhat flawed results. Nevertheless, combined with his observations of craters on the moon, solar flares, and the considerable mass of the solar system outside the Earth, the supposed Biblical view of the Earth as the center of a perfect universe began to erode.

Soon after Galileo's pronouncement, Johannes Kepler, using volumes of data collected by Tycho Brahe, astrologer for the Holy Roman Emperor, proposed that planets and moons moved in elliptical orbits rather than circular ones. Kepler's calculations had none of the imperfections of Galileo's calculations, providing compelling evidence that the heliocentric view of the solar system was correct. The religious authorities were forced to yield to physical reality. Science began to move beyond the authority of the church, even though Kepler himself was a devout Christian who attempted to glorify the Creator in all his scientific work.

The English physician, William Harvey, was the first to understand how blood circulates in the body. Before his treatise was published in 1628, it was believed that blood originated in the liver, was transported to the heart, and finally passed into the vessels of the body to be consumed as fuel. These ideas originated with Aristotle and Galen and were mixed with a fair amount of mysticism about the operation of the human body. In his book, *An Anatomical Study of the Motion of the Heart and of the Blood in Animals*, Harvey was able to show that blood was not continually manufactured but recirculated. He explained that blood flowed from the right side of the heart, through the lungs, back into the left side of the heart, into the arteries, and returned through the veins. Although he could not see the capillaries (the microscope had not yet been invented), he deduced that a capillary system had to exist. Harvey demonstrated that, although the human body is marvelously complex, it was possible for humans to understand its operation, that human physiology did not depend on continuous, miraculous intervention by the Creator.

Living at the end of the Renaissance and the birth of the Enlightenment, Harvey was the quintessential Enlightenment thinker. He rejected mystical theories based solely on philosophical tradition in favor of empiricism. Harvey described his methods as, "not from books but from dissections, not from the tenets of Philosophers but from the fabric of Nature" (www.lucknow.com). He saw that man's objectives were to harness nature for the benefit of mankind. Indeed, he wrote, "Civilization is simply a series of victories over nature" (www.greats2000.org). As a result, he was instrumental in promoting the doctrine of humanism, even though religion was still a very important part of most people's lives.

The scientific method was born at the transition from the Renaissance to the Enlightenment. Francis Bacon, an early seventeenth century English philosopher, developed the system of thought that replaced the guesswork and the citing of authorities of earlier times. In his *Novum Organum* (1620), Bacon proposed that scientists should observe a phenomenon, catalog the facts that result, and let the facts suggest their own order for understanding the phenomenon. Only after the suggested order was recognized, would the scientist make generalizations or propose hypotheses. Living at the birth of the Enlightenment, Bacon proposed the very Enlightenment-like ideas that science is the key to human progress and would eventually tame nature for humanity's benefit.

The time was ripe for an emphasis on empiricism. Europe was fatigued with religious disputes and wars that had Protestants and Catholics seeking to destroy each other in the name of revealed truth. These wars led to skepticism about the religious approach to knowledge. Perhaps this climate is what led Bacon to distinguish so clearly between divine revelation and direct observations as two distinct ways to approach truth, each valid in their own mutually exclusive spheres. Nevertheless, Bacon thoroughly believed that divine revelation led to a higher and more important body of knowledge (within the religious sphere).

The Enlightenment (Birth of Modernism)

The Enlightenment is extremely important because it gave birth to the modern way of thinking that still exists today (in spite of efforts by the post-modernists). The Enlightenment or Age of Reason began, according to most historians, when new hope filled war-ravaged Europe at the end of the Thirty Years War in 1648. However, the Renaissance had already set the stage with its rise in individualism, hope in humanity's potential, a system for scientific inquiry, and the Reformation which undercut the political and cultural authority of the Roman church. However, the Renaissance only set the stage with props. It was up to the Enlightenment thinkers to populate the stage, play the parts, and create the drama.

Prior to the Enlightenment, the prevailing worldview, which was established during the Middle Ages and recalibrated during the Reformation, placed God, primarily represented by Jesus Christ, at the highest point of all reality. The angels were God's "ministering spirits" and humans were made "a little lower than the angels." In this cosmology, God worked through nature and divine revelation to affect the affairs of mankind, the pinnacle of creation.

The Age of Reason turned this cosmology upside down, elevating humanity to the top. God still maintained importance, but He was constrained to compartments of reality assigned to Him by humans. This is not to say that Enlightenment thinkers were atheists. They believed that God was the Master Designer of the universe and that it is the duty of mankind to study nature so that they can understand the laws of God. However, the Bible as a source of understanding God's creative acts, was viewed as a symbol of the ecclesiastical authority that was being rejected in favor of human reasoning power. Human reasoning would usher in a period of unparalleled progress; this bold optimism led to utopian visions of the future of humanity, perhaps achieved with God's blessing and in accordance with His plan, but not necessarily by His power.

Scientists and theologians increasingly began to think in Baconian terms of two kinds of religion: natural and revealed. Natural religion included those fundamental principles that God had written on the hearts of humans, vis-à-vis, the existence of God and fundamental morality, principles that every man could access through his divinely provided reason. Revealed religion included specific Christian doctrines that had been derived from the Bible. This division almost seems planned for the ultimate purpose of preparing revealed religion for amputation while preserving basic morality that is so important to maintaining a civilized society.

Indeed, as the Enlightenment progressed, revealed religion fell into trouble and began to be identified with all that was wrong with the Dark Ages. The English empiricist, John Locke, wrote *The Reasonableness of Christianity* (1695), in which he set forth the thesis that Christianity is the most reasonable form of religion, provided one separates from it the idea of revealed truth. This resulted in the theological viewpoint of deism, which maintained a belief in God and the minimum of doctrine necessary for good morals. Combined with the Enlightenment emphasis on empiricism, deism resulted in the near elimination of the Bible as a source of knowledge other than moral principles. The subservience of the supernatural to the natural eventually led to the elimination of God in the thinking of most intellectuals later in the modern age.

Sir Isaac Newton has been called by some the Father of the Enlightenment. He discovered the mathematical relationships among objects of mass that affect their motion. Newton's three laws of motion are still used by physicists today. Every time the space shuttle is launched, the mathematical relationships Newton discovered are used to calculate the orbits. Newton also established the law of gravitation and the principles of optics.

It is difficult to overestimate the impact of Newton on both science and philosophy. He was the culmination of the Copernican revolution. Copernicus proposed the heliocentric hypothesis. By his observations, Galileo demonstrated that the model more closely resembled reality. Kepler refined the model. Finally, Newton explained the science behind the motion with mathematical precision. These men demystified the movement of the celestial bodies and dethroned the Earth as the focal point of Creation. In his *Principia* (1687), Newton set forth a universe as a machine-like entity, subject to mathematical laws that can be easily understood. Although Newton was a devout Christian, the unintended result of his work was the final demythification of nature; human study of nature did not need the aid of divine revelation. The Newtonian revolution eventually led to a worldview in which everything that happened was the natural outgrowth of particles responding in predictable ways to the laws of physics - a world without freewill or hope, a totally mechanistic world.

The Age of Reason came to an end approximately at the beginning of the nineteenth century. The religion of reason that had replaced Christianity in intellectual circles, was itself being subject to skepticism and relativism. British philosopher David Hume (1711 - 1776) and German philosopher, Immanuel Kant (1724 - 1804) set in motion a way of thinking that undermined many Enlightenment principles. Nevertheless, the scientific and technological progress begun with empirical thinking would continue. In spite of a prevailing belief in God among scientists, the pendulum was rapidly swinging from a superstitious view of God continuously pulling the marionette strings to one of humanistic self reliance. The nineteenth and twentieth centuries, the Age of Modernity, would see the pendulum reach its other peak, a time when naturalism and humanism would advance beyond the intellectuals and a scientific hypothesis on origins would become a philosophy of life masquerading as scientific fact.

Nineteenth Century Science - The Rise of Modern Evolutionary Theory

In spite of the fact that scientists are generally intellectuals, they tend to be more conservative than the leading thinkers of the times. For the most part, scientists are interested in pursuing science to the point of excluding anything that dilutes the effort. They are not interested in keeping up with the latest in philosophical theories. Therefore, they have tended to lag behind the wave of intellectual development that has been described in the preceding sections. It is not surprising then that scientists of the nineteenth century, although mechanistic in their view of nature, believed in God, at least as a first cause and to some extent as a personable Creator that sustains his Creation.

For example, in 1824, the first issue of the journal, *Zoological Journal of London*, stated (Denton 1986):

> The naturalist ... sees the beautiful connection that subsists throughout the whole scheme of animated nature. He traces ... a mutual depending that convinces him nothing is made in vain. He feels, too, that at the head of all this system of order and beauty, pre-eminent in the domain of his reason, stands Man ... the favoured creature of his Creator.

Similarly, the great zoologist, Louis Agassiz of Harvard wrote in 1857 (Denton 1986):

> [the living world] shows also premeditation, wisdom, greatness, prescience, omniscience, providence ... all these facts ... proclaim aloud the One God whom man may know, and natural history must, in good time become the analysis of the thoughts of the Creator of the Universe, as manifested in the animal and vegetable kingdoms, as well as in the inorganic world.

Biologists, therefore, felt little pressure to interpret observations of the natural world in the artificial framework of a totally mechanistic worldview. Consistent with both Scripture and their own observations, biologists saw the great panoply of life as discontinuous types of life, unlinked by transitional forms. Although there was clearly variation within the "types," the variation was bounded by the gaps between species (or some other type). Life forms were static (as we see them today too!), not evolving from one form to another. Indeed, the anatomist, Georges Cuvier, proposed that variation beyond the "types" would seem impossible, as reported by William Coleman (Denton 1986):

> The organism, being a functionally integrated whole, each part of which stood in close relation to every other part, could not, under pain of almost immediate extinction, depart significantly from the norms established for the species by the first anatomical rule. ... Transmutation by the accumulation of alterations, great or small, would thus be impossible.

Darwin was not the first to upset this almost too comfortable balance of science and theology. In 1830, Charles Lyell published, *Principles of Geology - Being an Attempt to Explain the Former Changes of the Earth's Surface by Reference to Causes Now in Operation*, which eventually caused revolutionary changes in geology and biology. Lyell's proposition, the theory of uniformitarianism, had four components, of which three are still considered useful.

(1) ***Natural laws are constant.*** Clearly, the study of nature could not proceed unless the physical laws of nature remain the same, subject to repeated testing. This seemingly reasonable axiom is not, however, always true if one believes that, at times, God intervenes in the affairs of this world in ways that seem to transcend physical laws. For example, during a battle with the Amorites, Joshua needed more daylight hours. He asked God to command the Sun to stand still. Joshua 10 verse 13 states, "*And the sun stood still, and the moon stayed, until the people had avenged themselves upon their enemies.*" Did God temporarily amend the laws that govern movement of heavenly bodies? Did he stop the

rotation of the Earth? If so, then this component of uniformitarianism is not unconditionally true. Some scientists today even theorize that the physical constants of the universe are slowly varying.

(2) ***Earth's geological features were caused by processes we observe taking place today*** or "the present is the key to the past." Again, this seemingly reasonable concept is not without limitations. The presence of discontinuities in the flow of nature would not be revealed by the present. Events such as the original creation of the world or the destruction of the world by flood would find no analogous events in the present.

(3) ***Geological changes are always slow, gradual, and steady.*** Even evolutionary geologists recognize the limitations of this principle. While there is a strong tendency among scientists to more readily accept hypotheses consistent with this component of uniformitarianism, most would recognize that catastrophic events have occurred in the Earth's past. While today's scientists believe that the sediments that comprise the Colorado Plateau were gradually laid down in a slow, steady process over millions of years, they are compelled to believe that massive sediments laid down near Mount St. Helens at its eruption in 1980 resulted from a cataclysmic event.

An examination of the Earth's surface from a uniformitarianism viewpoint leads one to believe extremely long time periods are necessary for its features to have developed. The sediments of the Grand Canyon, for example, must have taken many millions of years to deposit if laid down by currently slow rates of sediment deposition. Also, the erosion of the Canyon by the Colorado River must have taken a very long time given that river valleys known to mankind for thousands of years have not cut significantly into the rock beneath them. Thus, partially because of its challenge to Genesis, Lyell's uniformitarianism was not immediately accepted by scientists of his day. Nevertheless, the principle began to erode confidence in the Scriptures as a source of knowledge on origins.[1]

In December 1831, when Charles Darwin set sail on his famous voyage on *HMS Beagle,* he carried with him the newly published *Principles of Geology.* Darwin was an unpaid naturalist on the voyage, but he had no visions of greatness, no idea of the civilization-changing nature of the event. However, when the *Beagle* returned in 1836, Darwin had within his mind the beginning of an idea that would not only shake science but all of Western thought. His evolutionary ideas were not fully formed, and he had no concept of what mechanism would drive the evolutionary process. (A mechanism of acquired traits had been proposed by Jean-Baptiste Lamarck in 1809, but it lacked a sound scientific basis; modern genetic principles have subsequently proven it to be false.)

Darwin found an intellectual stimulus for defining the driving force behind evolution when, in 1838, he read *Essay on the Principle of Population* by Thomas Robert Malthus. Malthus described how plants and animals were under tremendous pressure of competition for sur-

[1] Interestingly, uniformitarianism and its impact on faith was predicted in II Peter 3: "Knowing this first, that there shall come in the last days scoffers, walking after their own lusts, and saying, Where is the promise of his coming? For since the fathers fell asleep, all things continue as they were from the beginning of the creation. For this they willingly are ignorant of, that by the word of God the heavens were of old, and the earth standing out of the water and in the water: Whereby the world that then was, being overflowed with water, perished." That is, in the last days, there will be people who do not believe the Earth was created by God and that He destroyed it by a flood, because they believe all things continue as they always have, i.e., uniformitarianism.

vival. More individuals are produced than can possibly survive. Darwin reasoned that under these conditions, small, favorable variations might be preserved in a species and unfavorable (that is for survival) would be destroyed. It wasn't until 1859, however, that Darwin published his theory in perhaps the most influential book ever written, other than the Bible, *On the Origin of Species by Means of Natural Selection or the Preservation of Favoured Races in the Struggle for Life.*[2]

The theory of evolution, as presented by Darwin, can be briefly summarized as follows. (Chapter 2 will provide a more complete discussion.) There is great variety among individuals of a species. Consider, for example, what breeders have done artificially maximizing the potential variety in dogs. The mechanisms responsible for the natural variations are blind to the needs of the organism, having produced variety from chance. Just as breeders artificially select characteristics desirable to the breeder and preserve them in the offspring, the pressures of survival naturally select traits favorable to survival. This is because as the traits enhance survival, they enhance the individual's potential to reproduce, thus passing on the trait to subsequent generations. This is called "natural selection" or "survival of the fittest." Accumulation of very small changes may eventually result in a new species. What has been described so far is known as the special theory of evolution. The special theory of evolution has been demonstrated to be true; cases of which have occurred within the lifetimes of scientists who observed them.

---

[2] Darwin was not the first to propose a theory of evolution. Indeed, evolutionary concepts can be traced back to the ancient Greeks. Compte de Buffon (1707 - 1788), Lamarck (1744 - 1829), and even Darwin's grandfather, Erasmus Darwin (1731 - 1802) all contributed evolutionary theories that helped pave the way for Darwin's. A contemporary of Darwin, Alfred Russell Wallace (1823 - 1923) proposed a nearly identical theory a few years earlier than Darwin, but he did not publish it in time to get credit for the theory.

What causes difficulty for those who believe God created life is the general theory of evolution. The general theory simply states that the special theory is responsible for evolution of all life from the first primitive life form. When Darwin first proposed the theory, most scientists refused to embrace it. The French biologist, Louis Agassiz (1807-1873) was the world's leading anatomist. Agassiz argued that the facts of biology do not support evolution. He believed in the biblical account of creation all his life. Physicists Michael Faraday (1791 - 1867), James Clerk Maxwell (1831 - 1879), and William Thomson (later Lord Kelvin; 1824 - 1907) never accepted Darwin's theory and defended the biblical account. Nevertheless, under a barrage of propaganda from Darwin and his so-called "bulldog," Thomas Huxley, the general theory of evolution became to be accepted as fact.

In science, theories are just that - concepts that need to either be confirmed (or more generally supported) by further experiment or else falsified. Scientists that propose hypotheses that prove to be incorrect are not usually judged too harshly, because the nature of science embodies this trial and error aspect. However, sometimes theories move too quickly to 'fact' status for reasons other than having received universal scientific confirmation. The theory of evolution is one such theory. The next section discusses how evolution became a philosophical worldview that is only challenged within the scientific community at the risk of excommunication from science.

From Scientific Speculation to Dogma

At the time Darwin wrote *Origin*, he could not point to any examples of natural selection having produced any biological change. Neither could he point to a continuum of life that supported gradual transition of life from simple to complex. Indeed, life must have appeared maddeningly discontinuous to him. The biologists of his day were typologists, believing that each distinct species always reproduced after its type. Indeed, Darwin wrote in his first edition of *Origin*:

> Why, if species have descended from other species by insensibly fine gradations, do we not everywhere see innumerable transitional forms? Why is not all nature in confusion instead of the species being, as we see them, well defined?

Furthermore, Darwin could find no evidence for gradual evolution in the fossil record. He wrote, "But as by this theory innumerable transitional forms must have existed why do we not find them embedded in countless numbers in the crust of the earth?" However, he attributed the lack of fossil evidence to insufficient investigation of fossil strata.

Therefore, Darwin was constrained to present his theory from logic and analogy, but most decidedly not from observation of natural selection in progress. Darwin reasoned that artificial selection by animal and plant breeders produced variation within a species. Could not the pressures of survival produce the same type of selection naturally, and over millions of years (granted him by Lyell and others), would not the variation be without bounds? Biologists of his day were not quick to accept Darwin's argument from analogy. They reasoned that artificial selection results from intelligent decisions, not random events. Also, when individuals with extreme variations resulting from artificial selection are returned to the wild, the striking characteristics vanish, and the offspring return to the original wild type, indicating that natural selection is a conservative force that inhibits striking change.

Darwin's theory was wonderfully elegant and insightful, but it was at best speculation borne of an intelligent and imaginative mind. Now there is nothing wrong with a theory being presented in a speculative framework. Great advances in science can be made this way. However, it is wrong when scientist abandon their requirement for evidence and accept philosophical speculations as established fact. Nevertheless, in the case of Darwinism, within 20 years, the theory of evolution became scientific dogma. Something unusual happened to cause scientists who pride themselves in their objectivity and empirical epistemology to embrace Darwin's proposition so thoroughly. Was it the discovery of new evidence unknown in 1859? Unfortunately, it was not, but there were four conditions that supported the unnatural acceptance of Darwinism.

First, Darwin's gradualism was elegantly consistent with the philosophy of 1859. Still very much under the influence of the Enlightenment, intellectuals were enamoured with the inevitability of continual human progress and the dream of ultimate human perfection. Natural selection provided the driving force that had supposedly improved life from some prototype life into humanity and would continue to operate until a utopian condition was achieved. Natural selection was also viewed as analogous to the competitiveness of the blossoming free market economy that was then driving great economic and social evolution. Therefore, the nineteenth century mind was much in tune with the concept of gradual improvement through evolution.

Secondly, Lyell's principle of uniformitarianism had produced great improvements in the science of geology in the 30 or so years since it was published. If the surface of the Earth evolved over very long time periods by processes currently being observed, could not a similar revolution in biology result in great advances in biological knowledge? The biologists certainly hoped so.

Thirdly, the scientific revolution in physics, from Copernicus to Newton, had eliminated the need for the supernatural to explain natural phenomena. Although the scientists were not quite ready to discard the concept of a Creator, they were willing to eliminate any dependence on Him to explain concepts that can be observed and tested. Biology, being the study of something much more special than atoms and crystals and stars, that is, of life itself, was to some extent, still dependent on a Creator. Biologists yearned for the same freedom that physicists enjoyed and for the same degree of advancement and rigor in their field. Therefore, they were intellectually prepared for a grand theory of biology that would grant the same rights and privileges that the physicists enjoyed.

Finally, and perhaps ultimately most importantly, biologists (and most of the rest of humanity) wanted freedom from accountability to a God who limited the lifestyles of His creatures. Julian Huxley (1887 - 1975), the eminent evolutionist and grandson of Darwin's colleague, Thomas Huxley (1825 - 1895), explicitly stated it: "We all jumped at the *Origin of Species* because the idea of God interfered with our sexual mores."[3] For this reason and the others stated above, people have been eager to reject a knowledge of God. Such willful rejection of obvious truth can lead to foggy thinking. The Bible describes the process, "*And even as they did not like to retain God in their knowledge, God gave them over to a reprobate mind to do those things which are not convenient.*" Romans 1:28.

Eventually, scientists overlooked the lack of evidence for evolution and the lack of a mechanism that would explain how one species could change into another and simply stated, "Evolution has to be true because here we are. There is no other acceptable explanation." Had anything changed in the facts of nature from 1859 to 1900 or even in mankind's understanding of nature? Of course, knowledge advanced, but nothing arose to give evidence to support Darwin's theory. Scientists, and eventually the public, simply moved from one philosophical worldview to another. Prior to Darwin, nature was interpreted from a viewpoint influenced by religion. After Darwin, nature was interpreted with the *a priori* assumption that God was irrelevant in nature. Naturalism became the philosophy of science.

[3]The author regrets that the source for this quotation has been lost. It was used several years ago in a presentation from which the current usage was borrowed. However, the presentation slides did not indicate a source. The demanding reader may elect to ignore the quotation.

Nowadays, biological evolution is an established fact in most educated circles - no longer a scientific theory. In 1959, at the centennial celebration of the publication of *Origin*, Julian Huxley was the honored speaker (Johnson 1993):

> In the evolutionary pattern of thought there is no longer either need or room for the supernatural. The earth was not created, it evolved. So did all the animals and plants that inhabit it, including our human selves, mind, soul as well as brain and body. So did religion ....
>
> Finally, the evolutionary vision is enabling us to discern, however incompletely, the lineaments of the new religion that we can be sure will arise to serve the needs of the coming era.

Today, every scientific journal, textbook, and academic discussion assumes the truth of Darwinism and never challenges it. This creates a powerful bias against ever challenging the theory - a condition that should be intolerable in science, for science thrives on challenging its own theories. However, the facts today are largely the same as they were in 1859. As we shall see in Chapter 2 and others, there is no evidence for evolution, and no mechanism for transmutation of species has been established.

Notice again the lack of scientific treatment of evolution by Julian Huxley's 1959 speech: "The first point to make about Darwin's theory is that it is no longer a theory but a fact ... Darwinism has come of age so to speak. We are no longer having to bother about establishing the fact of evolution." Richard Dawkins, in *The Selfish Gene,* writes (Britannica 1999): "The theory is about as much in doubt as the earth goes round the sun."

A theory as thoroughly embraced as fact as is evolution cannot be falsified in the minds of its adherents. The science of biological evolution has become a search for confirming evidence. However, there is a difference between experimenting to test a theory and observing to confirm the only theory allowed. The latter is not science but pseudoscience. One can always find evidence to believe a preconceived idea. It is easy to overlook seeming falsifications as poor experimental method or to create a modification to the accepted theory to accommodate the errant observations. The same thing happened with Marxism and Freudism; both arose at approximately the same time as Darwinism, both were promoted as scientific, both produced theories that could never be falsified, both explained everything within their spheres of influence, and both are now largely discredited.

Scientists have created an environment in which they have unwittingly limited their range of inquiry. Only theories consistent with the philosophical bias of naturalism are acceptable for investigation. But what if naturalism is not true?

A logical question one might ask is whether science has found, in the years since Darwin, evidence supporting biological evolution, whether there is a logical explanation for the scientists seeming to abandon their own principles of science? The next chapter will deal with the proofs of evolution.

Two

Charles Darwin

His Theory and its "Proofs"

Most mothers are convinced that their boys will accomplish great things once properly prepared for the world. However, when on February 12, 1809, Susannah Wedgwood Darwin examined her newborn son, she could not possibly have imagined the degree to which baby Charles Robert Darwin would influence the world. Indeed, most people today have little concept on how different their lives have been because of Charles Darwin and his theory.

Charles was born to a wealthy and intellectual family in Shrewsbury, England. His mother was the daughter of Josiah Wedgwood, founder of the famous Wedgwood pottery. His father, a well known physician, was the son of Erasmus Darwin, also an illustrious physician and author of *Zoonomia*, an early work on biological evolution.

After attending a Dickens-like grammar school in which virtually the entire curriculum was in Greek or Latin, young Charles was sent to Edinburgh University to study medicine. Charles did not do well in medical school. So after two years, his father arranged for him to attend Christ's College at Cambridge to study for the ministry. Darwin's curriculum at Cambridge consisted of three general subjects: classics, mathematics, and theology. Again, Darwin was a poor student, but two of the books he was required to read in his theology course were William Paley's *Evidences of Christianity* and *Moral and Political Philosophy*. He enjoyed Paley so much that he also read *Natural Theology*, which, as we discussed in the first chapter, addresses evidence for a designer of the world.

While at Cambridge, Charles enjoyed the influence of two outstanding mentors: Adam Sedgwick, a geologist, and John Henslow, a biologist. Henslow taught Darwin to be a meticulous and painstaking observer of natural phenomena and collector of specimens. After Darwin's graduation from Cambridge in 1831, Professor Henslow arranged for the 22-year-old Darwin to be taken aboard HMS *Beagle* as an unpaid naturalist. *Beagle* was to be sent on a five-year scientific expedition around the world.

Beagle was under the command of Captain Robert FitzRoy, a religious man who believed in the literal interpretation of the Bible. Captain FitzRoy held services every Sunday for five years and seemed to have some positive influence on the nominally religious Darwin. Ironically, one of FitzRoy's objectives for the voyage was to substantiate the Genesis account of creation and world-wide flood.

During the voyage, Darwin studied geology, botany, and zoology as they circumnavigated the globe. Because Darwin knew little of geology, he studied the newly-released *Principles of Geology* by Charles Lyell, which presented uniformitarianism, a theory requiring millions of years for the Earth to have evolved its features. Darwin made copious notes on his observations. When he returned to London, he may have puzzled over the things he had seen, but as yet, he had no concept of a theory of biological evolution. However, the seeds of the idea were present, and the data were in his notebooks. He also began to read extensively on the species question. Over the following years he formulated his so-called "dangerous idea" and finally published it in 1859. The book contributed in a major way to the secularization of society.

Although *Origin of Species* was not especially well written and contrary to the prevailing scientific views of the time (even Lyell initially refused to accept it), it was an instant best seller. Within a few days, the entire printing was sold. The second edition was printed two months later in January 1860. The sixth and final edition was printed in 1872. Because each succeeding edition attempted to address challenges to the theory, many consider the first edition to be the best. Interestingly, the term "survival of the fittest" was borrowed from Herbert Spencer and didn't appear in *Origin* until the fifth edition. The term "evolution" did not appear until the sixth edition.

Some historians have made the case that Darwin was less than honorable about borrowing other men's ideas. Others have nearly deified him. Whatever the truth, today, Darwin gets credit for the book that "shook the world." In order to understand the book's great impact and the validity of Darwinism today, we will need an understanding of what is in the book. Amazingly, in spite of advances in biology, especially in genetics, *Origin* is still one of the best treatises on the subject of biological evolution.

The Theory of Biological Evolution

The Galapagos Archipelago is comprised of 13 small volcanic islands west of Ecuador. Although Darwin explored the coasts of Patagonia as well, it is the Galapagos that are most closely associated with his name. Darwin noted that there were interesting variations among the various plant and animal species across the different islands and the mainland. It was as if each isolated population of finches, for example, had departed in morphology from the original breeding stock in accordance with varying environmental influences on the islands. In fact, with careful study, an anatomist could arrange the finches in a morphological sequence suggesting common descent.

To Darwin, this suggested that possibly the species may not be fixed: "...the zoology of Archipelagos ...will be well worth examining; for such facts would undermine the stability of Species," he wrote in *Origin*. Again, he wrote:

> Seeing this gradation and diversity of structure in one small, intimately related group of birds, one might really fancy that from an original paucity of birds in this archipelago, one species had been taken and modified for different ends.

Although Darwin had no evidence that the various species had indeed descended from a common ancestor, he certainly could not be blamed for hypothesizing it.[4] The evolution of the finches seemed more likely to him than the idea that each finch variety or species was a specific act of creation by God. In fact, Darwin was correct. The only reason variation such as demonstrated by the finches was thought to be contrary to Scripture was because scientists in his day had too rigid an interpretation of the fixity of species. Genesis states that the animals would reproduce "after his kind." Genesis does not define what a "kind" is. The original created kinds or types may well have had a richness of genetic potential to accommodate large degrees in variation within the bounds of the kind.

Darwin saw many marvelous things that he recorded in his notes, but of all, the variation under nature, as he put it, was the primary observation that led to the theory of evolution. He saw nothing that suggested the major types of nature had been crossed through gradual change. He saw nothing to suggest that life had originated from non-life. He saw nothing indicating that man had originated from an ape-like ancestor. If Darwin had theorized based on what he had observed, he would have made a significant contribution to science - the special theory of evolution (or microevolution) in which a species or some other type varies under environmental pressures within the bounds of the type. However, he went well beyond observation to produce fanciful speculations. His theory can be summarized into five points[5] :

---

[4] It wasn't until 1977 that natural selection among the Galapagos finches was demonstrated by observation. The beak sizes in the population of finches oscillated from year to year depending on rainfall and the resulting seed sizes ["Oscillating Selection on Darwin's Finches" Nature 327, p. 511, 1987, as reported by Johnson (1993)].

[5] Except where noted, the accompanying quotations are all from *Origin*, 1st edition.

1. The species are not immutable; they undergo descent with modification.

> I am fully convinced that species are not immutable; but that those belonging to what are called the same genera are lineal descendents of some other and generally extinct species.

2. The process by which the species vary under natural selection[6] provides the diversity of all life that we see. That is, all life descended from a common, more primitive ancestor or perhaps a few primitive ancestors.

> Therefore, I should infer from analogy that probably all the organic beings which have ever lived on this earth have descended from some one primordial form, into which life was first breathed.

3. The variations in species (or in our day, the mutations) upon which natural selection acts to fix the new trait, must be randomly generated, not directed.[7]

> It may be said that natural selection is daily and hourly scrutinizing throughout the world, every variation, even the slightest, rejecting that which is bad, preserving and adding up all that is good; silently and insensibly working whenever and wherever opportunity offers, at the improvement of each organic being in

---

[6] Natural selection was defined in the first chapter. Simply, it is the principle of survival of the fittest.

[7] The author could not locate, in the time allocated to the task, an explicit quote from Origin. The concept of random variation is implicit in the quoted text.

> relation to its organic and inorganic conditions of life. We see nothing of these slow changes in progress, until the hand of time has marked the long lapses of ages, and then so imperfect is our view into long past geological ages, that we see that the forms of life are now different from what they formerly were.

4. The steps from one species to another, through the generations, must be inscrutably small; otherwise, the integrated self-adapted nature of living creatures would be disrupted. Large changes would have required numerous, simultaneous modifications such that the probability becomes vanishingly small and suggestive of a miracle or act of creation.

> As natural selection acts solely by accumulating slight, successive, favourable variations, it can produce no great or sudden modifications; it can act only by short and slow steps.

5. The process of natural selection has a corollary in the theory of extinction. When competing for survival, the parent species would be at a disadvantage compared to the new species. The parent would lose the competition and become extinct. Extinction contributes to the discontinuity among the types observed in nature.

> Extinction ... has played an important part in defining and widening the intervals between the several groups in each class. We may thus account for the distinctness of whole classes from each other - for instance, of birds from all other

> vertebrate animals - by the belief that many ancient forms of life have been utterly lost[8] ....

Darwin presented these concepts in the fourteen chapters of the book. The first five chapters primarily address microevolution: analogy with artificial selection, the natural variation existing in nature, the pressures of competing for survival, the principle of natural selection, and principles of its expression in nature. These initial chapters primarily address the first of the five points of evolution, that the species are not immutable. It was necessary for Darwin to firmly establish this principle if he were to make any headway at all on the general theory, sometimes called macroevolution, which was largely the subject of the subsequent chapters.

Chapters 6 through 8 deal with difficulties with the theory that Darwin honestly recognized. Chapters 9 and 10 deal with geology and paleontology. Chapters 11 and 12 address geographical variation which was, as we have seen, one of the observations compelling Darwin to abandon the immutability of the species. Chapter 13 speaks to morphology, homologies[9], embryology, taxonomy, and rudimentary organs. We will deal with many of the issues Darwin raised in Chapters 6 through 13 in later sections of this chapter.

Finally, Chapter 14 recapitulates the arguments of the book and suggests that the principles of evolution can be extrapolated to explain the existence of all life forms as originating from one primordial ancestor. Darwin came short of suggesting that life originated from non-life or

---

[8] *Origin* 6th Edition as reported by Denton (1985), page 136.

[9] Homologies are structures that are similar across many species and suggest common origin to some scientists.

that humans occupy the top of the evolutionary ladder; however, one would have to be dull-witted indeed to miss the implication.

We will now examine some of the evidences for evolution from a modern perspective, that is, arguments for the efficacy of the theory that can be found in today's college text books.

Modern Arguments for Darwin's Theory

This review of the evidence for biological evolution is not intended to be exhaustive. Nevertheless, it encompasses what are generally thought to be the strongest arguments by the theory's proponents. First, one must recognize that there is a certain symmetry in nature by the way plants and animals are categorized, suggestive of common origin. Second, there are characteristics that are common across many species, a condition that could suggest common origin. Next, the fossil record should provide a clue as to whether the discontinuity we see in current life forms is violated among the extinct and historic life forms. Fourth, a leading evolutionist has provided a list of "proofs" of the certainty of natural selection and biological evolution. Finally, the study of biochemistry and molecular biology offers exciting new ways to look for evidence of descent.

Systematics

Humans have an innate tendency to organize complex information, to classify it, so that general principles reveal themselves. We organize stars into constellations, library books into catalog systems, and plants and animals into taxonomic categories. Aristotle was the first known taxonomist. He attempted, mostly by compara-

A Typical Modern Classification Scheme

Taxonomic element	Example
Kingdom	Animalia
Phylum	Chordata
Class	Mammalia
Order	Carnivora
Family	Canidae
Genus	Canis
Species	Lupus

tive anatomy, to organize all living things into categories based on what they are "in essence." Carolus Linnaeus (1707 - 1778) was the first modern taxonomist. It was Linnaeus who devised the binomial system of nomenclature by which living entities were first classed by their general "type" and then by the more defining characteristic, as in *Canis lupus* where *Canis* is the genus name or the general type of entity, in this case dog-like mammals (including dogs and wolfs) and *lupus* is the species name or the specific type known as the wolf.

Aristotle and Linnaeus, followed by later taxonomists such as Georges Cuvier (1769 - 1832) and their currently living counterparts, have all faced the same difficulty: what characteristics do taxonomists use to categorize the various biological types? What is the true pattern of nature? Although over the years, biologists have applied increasingly objective and quantitative criteria, classification remains very much a subjective endeavor. Therefore, evolutionary biases have affected classification schemes to some extent. While Linnaeus believed the biblical account of creation, Ernst Haeckel (1834 - 1919) proposed the first scheme based on evolutionary principles.

Interestingly, all classifications of nature demonstrate a hierarchical system. The early taxonomists who were creationists believed that at the base of every categorical element was an archetype, a fundamental model in the mind of God, that guided the creation of the diverse creatures based on it. The evolutionists believe that at the branches where diversity occurs reside the common ancestors. Regardless of the bias of the taxonomist, the classification system always results in groupings in which every group is either totally inclusive or totally exclusive of every other group. There is no continuum or blending of natural groups, no sequence of increasingly complex life forms.

Darwin claimed that classification provides such decisive evidence for evolution that he would adopt the theory on its basis alone. Modern biologists also agree that taxonomy is strongly suggestive of evolution (when viewed in its evolutionary context, taxonomy is often called systematics). In spite of the natural typological revelation in the classifications, it is easy to see why one biased toward evolution sees common origin in the order and structure of nature. It is also easy to see why creationists see discrete types reproducing after their kind.

One of the more recent developments in taxonomy is the method of cladistic analysis. Cladism assumes no *a priori* bias on how the classification should develop. In cladism, all life forms are graphically portrayed in a branching set of lines resembling a tree. At the branch points or nodes are characteristics in common (known as homologies) with all branches emanating from that node. Because cladism identifies homologies based on morphology, not presumed common ancestors, there is debate among biologists of this method's value. The resulting cladistic diagrams show that all the members of a grouping are roughly equidistant (morphologically) from the node. This occurs at every hierarchical level. For example, the wolf, dingo, coyote, dog, fox, and jackal are all equally diverse from the homological features of the family *Canidae*. Similarly, the *Canidae*, bears, cats, seals, walruses, and five other families of carnivores are equidistant from the homologies of the Order *Carnivora*. There are not any linear sequences from a common ancestor through any of the various species under *Carnivora*.

Evolutionists must be frustrated indeed that cladism (or any classification method) results in such a distinct and hierarchical order of nature. Professor Keith Thompson echoes the concerns of many modern taxonomists (Denton 1985):

> No one needs reminding that we are well into a revolutionary phase in the study of evolution, systematics, and the interrelationships of organisms. ...[T]o the thesis of Darwinian evolution ... has been added a new cladistic antithesis which says that the search for ancestors is a fool's errand .... It is a change in approach that is not easy to accept for, in a sense, it runs counter to what we have all been taught.

Therefore, we see that with the vision of evolutionary bias, the order of nature may suggest descent with modification; however, there is nothing compelling in systematics that drives one away from belief in created types. Indeed, the typological discreteness of nature may be the most compelling evidence there is for creation by the Word of God.

Homologies

It is a marvel of nature that the forelimbs of the rat, dog, horse, bat, whale, mole, porpoise, man, and many other animals all have the same general bone structure known as pentadactyl design. Although the specific design is adapted to the various differences in use, they all have a humerous (upper arm in a human), radius and ulna (bones of the forearm in a human), carpals (wrist in a human), metacarpals (upper part of the hand in a human), and phalanges (fingers in a human). Such a similarity among the species is known as a homology when the similar features are thought to have been derived through descent from a common evolutionary ancestor. When structures have a functional similarity based merely on similar use, the commonality is known as an analogy. Biologists do not always find it easy to determine which similarities are homologies and which are analogies.

Darwin believed that homology is evidence for descent with modification. He wrote in *Origin*:

> What can be more curious than that the hand of a man, formed for grasping, that of a mole for digging, the leg of a horse, the paddle of a porpoise, and the wing of the bat should all be constructed on the same pattern, and should include similar bones, in the same relative positions. ... [I]s it not powerfully suggestive of true relationship, of inheritance from a common ancestor?

Evolution textbooks today still cite homology as an aspect of nature that is coercive evidence of evolution. In fact, it is possible that had there been no homologies in nature, there would never have been a theory of evolution. The questions remains, however, does the homology argument really compel a person to accept evolution or can another explanation also satisfactorily explain this curiosity?

Had evolution produced homologies, one would expect to see the homology originate in similar genes from the various species and become manifest through differentiation in the embryo of the same general tissues. However, as described by Denton, mutation of plieotropic[10] genes have been shown to affect homologous structures in ways suggesting that nonhomologous genes are affecting the homologies. That is, the genetic origin of a homology in one species results from different, nonhomologous genes in another species. Also, homologies can seldom be traced back to similar tissues in the embryo. Therefore, homologies appear among the species by ways suggestive of a non-evolutionary origin.

Another problem with homologies is that as scientists continue to study them, they find that many turn out to be analogies. If the current trend continues, biologists may find that one of the mainstays of evolutionary theory evaporates, under scrutiny, into analogies. After all, it is the homologies that taxonomists use to define the classes of lifeforms into discrete groups. It is the homologies that reveal the typological nature of life. Therefore, the very possibility of classifying life is evidence against evolution.

Do homologous structures really demand descent from a common ancestor? If one believes in evolution as a

[10] Plieotropism is the phenomenon of a single gene affecting characteristics of widely dissimilar tissues, for example, both air sacs and downy feathers in birds. The genetic and embryological discussion supporting this paragraph are technically beyond the level of this presentation. The reader is referred to Denton, Chapter 7.

philosophical worldview, then clearly one will see homologies suggesting evolution. If one believes in a Creator of all life, then one sees intelligent use of a good design that has been adopted and adapted for a variety of purposes.

Fossil Record

The Earth's crust is comprised of three principal types of rocks, classified by their origin. Igneous rocks, such as granite, are those derived from volcanic action and have solidified from the magma. Sedimentary rocks, such as limestone or shale, are those resulting from deposits of sediment, primarily by water. Metamorphic rocks, such as marble, are those that were originally igneous or sedimentary but have undergone major changes from heat and pressure. Sedimentary rocks constitute only 5 percent of the Earth's crust but they cover 75 percent of the Earth's surface. It is generally within the top 25 percent of sedimentary rocks, those nearest the surface, that fossils are found.

Fossiliferous rocks are further classified into eras and periods, based on the types of fossils found in them. The ages of the eras were, for a hundred years, determined relatively by study of stratigraphy and semiquantitatively by the presumed evolutionary ages of the fossils contained in them. The advent of radiometric dating has caused some recalibration of the ages of the eras and periods; however, index fossils[11] remain the primary method of dating sediments. Because it is not practical to radioactively date all samples, a certain amount of circular reasoning exists since fossils are

[11] Index fossils are those that are thought to represent extinct lifeforms that lived only a short period of time but that were widespread across the surface of the Earth. Paleontologists and stratigraphers use index fossils to correlate sedimentary layers that are not contiguous.

often dated by the sediment age and the sediment by the fossil age.

Darwin was somewhat perplexed as to why the fossil record did not support his theory. In *Origin*, he lamented, "But, as by this theory innumerable transitional forms must have existed, why do we not find them embedded in countless numbers in the crust of the earth?" He blamed the lack of support on the imperfection of the geological record and believed that future paleontologists would find the missing links.

Ever since Darwin, those future paleontologists have been busy indeed looking for evidence to support the general theory of evolution. From reading the popular articles, one would conclude that the fossil evidence is overwhelming in favor of Darwin's theory. Yet the same excuses are being offered today: "The fossil record is incomplete," states *Encyclopedia Britannica* on its web-based publication in April 2000 and then goes on to give the same old tired, overworked examples (we will address the horse fossils in Chapter 7). Leading evolutionary science educator, Douglas Futuyma declares, "The fossil record includes many gaps among quite different kinds of organisms, as well as gaps between possible ancestors and descendants. Such gaps can be explained by the incompleteness of the fossil record" (Futuyma 1998)[12].

In moments of honesty, paleontologists admit the lack of evidence for transitional forms. Professor Stephen

[12] Dr. Futuyma goes on to say, "But the fossil record also includes examples of gradations from apparently ancestral organisms to quite different descendants." In context Dr. Futuyma is not highlighting the weakness of the fossil record; however, the quote illustrates how evolutionists continue to be defensive about the lack of transitional forms.

Jay Gould[13] of Harvard University writes (Johnson 1993):

> The history of most fossil species includes two features particularly inconsistent with gradualism:
>
> 1. Stasis. Most species exhibit no directional change during their tenure on earth. They appear in the fossil record looking pretty much the same as when they disappear; morphological change is usually limited and directionless.
>
> 2. Sudden appearance. In any local area, a species does not arise gradually by the steady transformation of its ancestors; it appears all at once and "fully formed."

Again Gould writes:

> We can tell tales of improvement for some groups, but in honest moments, we must admit that the history of complex life is more a story of multifarious variation about a set of basic designs than a saga of accumulating evidence.... [T]he failure to find a clear vector of progress in life's history [is] the most puzzling fact of the fossil record.

---

[13] Gould and his colleague Niles Eldredge are at odds with most biologists over their proposal called Punctuated Equilibrium. Their criticism of the fossil record was offered to support a theory under attack by the majority. Their defensive posture could be the reason for their willingness to be so honest about the facts. Gould and Eldredge are vigorous evolutionists.

In reality, the fossil record shows major groups appearing suddenly, unlinked to other groups by transitional forms. Dr. Eldredge agrees: "We paleontologists have said that the history of life supports [the story of gradual adaptive change], all the while really knowing that it does not" (Johnson 1993). The earlier representatives of a class or family, although often different than today's representatives, clearly have all the homologies that distinguish the group. Identifying them as primitive forms of their modern counterparts only reveals a bias toward evolution.

So why do textbooks and the popular press indicate that the fossil evidence is so compelling? If biologists are convinced that descent with modification is a reality, then the paleontologists job is to establish the certainty of the theory with so-called evidence. Under such a scheme, false positives are ensured. Darwin claimed there were gaps; paleontologists have filled those gaps on demand, whether or not the evidence objectively fit. Some of the "proofs" offered are the horse sequence, *Archaeopteryx*, and *Therapsida*[14]. However weak these evidences may be, their small number, rather than confirming evolution, draws attention to the consistent pattern of stasis, sudden appearance, and discontinuity. Rather than looking for confirmations, scientists should be willing to test the theory and let it fail if it does not withstand scrutiny.

One of the aspects of the fossil record most perplexing to scientists is the so-called Cambrian Explosion. The Precambrian rocks contain only a few microscopic examples of life. Yet the Cambrian layers contain nearly

---

[14] The horse sequence is not as smooth and clean as the textbooks lead one to believe. *Archaeopteryx* is a mosaic fossil creature that is clearly reptilian but has feathered wings. *Therapsida* is an entire order of fossils that appear to progress from reptile to mammal. *Therapsida* will be discussed later in this chapter. The other examples will be discussed in subsequent chapters.

all the invertebrate phyla, fully formed and differentiated, without a hint of their evolutionary ancestors. Richard Dawkins, that stalwart Darwinist, writes: "It is though they were just planted there, without any evolutionary history" (Johnson 1993). Although not so dramatically appearing from nothing, every vertebrate phylum also eventually appears in the geologic column suddenly, without ancestors.

While not everything in the fossil record is easily explainable under either an evolutionary scenario or a creationist scenario, there is nothing that compels the observer to select evolution as the most logical explanation. In fact, the fossils are clearly in contradiction to Darwinist evolution. Some fossil difficulties for creationists will be addressed in Chapter 7, including human fossils, *Archaeopteryx*, the horse sequence, and dinosaurs.

Gould's Proofs

Often evolutionists cease citing evidence for evolution and begin citing proofs. In April of 2000, the *Encyclopedia Britannica* web-based article, "The Evidence for Evolution" stated, "Evolutionists are no longer concerned with obtaining evidence to support the fact of evolution, but rather with what sorts of knowledge can be obtained from different sources of evidence." There is hardly an evolutionist more adamant that evolution is a fact than Professor Gould. In an article, "Evolution as Fact and Theory," in the book, *Hen's Teeth and Horse's Toes*, Gould offers three "proofs" of evolution.

Gould's first proof is the existence of microevolution. Like Darwin, he cites the successes of breeders employing artificial selection. Unlike Darwin, who had no natural examples, Gould cites the case of industrial melanism. Gould calls the examples of microevolution, "evolution in action," but are they?

With intelligent intervention, breeders are able to make rapid changes in characteristics of the species under consideration. However, breeders always reach a genetic limit on the amount of change possible. No matter the selective pressure, even after many hundreds of years of breeding, the change stops cold, and the ultra-bred individuals often become sterile or sickly. All breeders are doing is expressing genetic potential that already exists in the breeding stock; they are creating no new characteristics, and they do not create new types.

However, industrial melanism is a phenomenon that represents the first documented example of natural selection actually occurring. Strangely enough, this first evidence for natural selection was discovered in the early 1950s, 100 years after *Origin*. Bernard Kettlewell noted that back in Darwin's time in England, the peppered moth was light colored and was well camouflaged against the trees and rocks of the area. As pollution began to darken the trees and rocks in industrial areas, the moths became more obvious targets for birds. Today, most peppered moths are dark in industrial areas but lighter in more clean areas.

Is this evidence for natural selection? Certainly. Moths carrying dark genes were preferentially selected to reproduce, because they weren't eaten as rapidly. Lighter moths had lesser opportunities to reproduce. However, the moth population always carried genes for both light and dark coloration. Industrial melanism acted to change the ratio of dark to light moths, but created no new characteristics. Since the discovery of natural selection in peppered moths, scientists have observed several other cases, including changes in beak size in Galapagos finches that track with wet and dry seasons.

Gould is an unusually brilliant scientist and philosopher, but it is difficult to understand why he believes that natural selection is a proof of evolution. Natural selection is almost a self-evident principle, but its only relationship to evolution is that Darwin, Dawkins, and others require it to operate for evolution to take place. Natural selection is a necessary condition for Darwinism to take place, but it is obviously not a sufficient condition. Had natural selection been proven to not take place, it would have been a victory for the creationists, but establishing that it does occur provides nothing to prove evolution. The last section of this chapter will speak more to natural selection.

Gould's second proof of evolution is the argument from imperfection. If God is the infinitely intelligent designer, why does nature show imperfection in its design. Gould cites the homology of the pentadactyl design as inefficient for the various functions it is called upon to perform in the bat, porpoise, and human. God would have optimized the design for each different application, but the homology is indicative of a common ancestor, because natural selection must act on existing structures. Gould made this same point in his 1980 book, *The Panda's Thumb*, in which he argues that the Panda's thumb is poorly designed for its purpose.

This very common argument is first of all a theological one without the backing of scientific analysis. It presumes to know what would be in the mind of an omnipotent and omniscient designer. The argument is also sufficiently arrogant to presume to know what design is truly best. It fails to account for the fact that designers often do not optimize engineering excellence but some other endpoint that would never occur to the creature that was the object of the design. But most importantly, the argument is merely a smokescreen for the lack of evidence for Darwinism. It does not substitute for scientific evidence supporting evolution.

Gould's third and final proof is that in spite of the lack of a "clear vector of progress" in the fossil record, there are some examples of transitional forms in the reptile-to-mammal transformation and the hominid-to-human transformation. The current speculation is that ancient ancestral fish evolved to amphibians, which evolved to reptiles, which evolved to mammals, which eventually produced primitive simians that evolved to humans in one branch and apes in another.

The fish to amphibians transition has been described in fanciful stories about what might have happened, but there is no solid evidence that the transition occurred. Fish learning to breathe and move on land, while simultaneously evolving the strange reproductive cycle of amphibians, is almost beyond belief. Similarly, the amphibian to reptile transition is fraught with difficulties such as changing from an egg-to-larva-to-adult reproduction in amphibians back to eggs in reptiles. No transitional fossils support the speculation.

However, the case is slightly different in the reptile-to-mammal transition. Reptiles have multi-bone lower jaws. Mammals have single bone jaws. There is an entire order of fossil species, *Therapsida*, that have jaws intermediate between reptiles and mammals. Some species have many bones in the jaw, and some have only one. If this characteristic were what truly represents the "essence" of reptiles or mammals, then paleontologists might have some evidence, but there are many features more characteristic of mammals and reptiles than jaw bone structure, such as the reproductive system. Convergence in skeletal features does not necessarily imply an evolutionary transition.

Even among the *Therapsida,* one cannot align the species in a single progression from reptile to mammal. A line of descent can only be established by mixing species of various subgroups. Also, there is no species at one end that is clearly mammal and no species at the other end that is clearly reptile. Finally, the argument that some of the jaw bones in the reptile species were transformed into the hammer and anvil in the mammalian ear defies the imagination.

Therapsida is the best argument that paleontologists have for transition in the vertebrate sequence. However, as usual, we see that the evidence is far from compelling. It represents a search for confirming evidence in place of investigation to test a theory. Were *Therapsida* not so strikingly an isolated case in the search for transitional forms, evolutionists might be forgiven for so often referring to it as providing "proof." We will consider the case of the hominids in Chapter 7.

Protein Sequences

Darwin had no possibility of understanding molecular biology; however, evolutionists today claim that the molecular evidence is confirming Darwinism. The discussion of how evolutionists use molecular data is too complex for this book. The discussion that follows addresses the subject in generalities. An interested reader should consult Michael Denton, *Evolution: A Theory in Crisis,* for a very readable presentation of the basic concepts.

Cells use the very complex molecule DNA to synthesize proteins. DNA is like an extremely complicated set of instructions, containing more densely packed information than could be stored in any digital mass storage device of today. The information stored in DNA is the genetic information inherited from an organism's parents. Proteins themselves are very complex molecules that are composed of long chains of amino acids. There are 20 types of amino acids. Proteins are distinguished by the order and number of amino acids in the chain. The varying structures of the proteins are one of the distinguishing characteristics of species within the various higher tiered classifications.

Cytochrome C is a protein that has been widely studied and is found in a wide variety of organisms. This protein has about 100 amino acids linked together; however, the sequence is different from species to species. By studying the amino acid sequence of cytochrome C, biochemists are able to classify organisms into groups. Organisms with very similar amino acid sequences can be grouped as related in some fashion. The question is, does the grouping represent anything that is fundamentally alike among the organisms? The astounding answer, although circumstantially supported, is that amino acid sequencing reveals a quantitative method to classify all life that produces very similar results to methods based on morphological considerations.

Just as evolutionists have attempted to use morphology to assemble presumed chains of descent, so have they done with amino acid sequences. However, the conclusion is the same. Molecular biology reveals a living world classed by types that are separate and distinct. Apparently even more sure than morphology, the molecular evidence shows the classifications of living creatures to be equidistant, biochemically, from every other classification at the same level. There is no molecular continuum of life.

Again, we are faced with very bright scientists looking at this same data and concluding support for evolution. However, relationship is not equivalent to ancestry. A biochemical or morphological relationship cannot confirm Darwinism, because Darwinism does not predict the relationship at all. The relationship is merely consistent with Darwinism (provided one is not troubled by lack of transitional forms in the fossil record). Therefore, when one looks for proofs of a theory, as opposed to testing a theory, the logical response to consistency is to assume proof. In reality, the biochemical evidence strongly suggests typology, a prediction of creationism.

We have examined some of the most important evidences for evolution that have been proposed by the theory's apologists. There are other specific examples for evolution that could have been discussed, but most fall under the general issues already discussed above. An excellent, but rather defensive, book on evolutionary biology is that by Douglas Futuyma of State University of New York, *Evolutionary Biology*. This book is the standard textbook of today and offers more arguments and some interesting specific cases. It even contains an appendix on how to respond to creationists. We now step back and look at natural selection itself and its partner, mutation, to see if there is any creative power in them.

Are Mutations and Natural Selection Creative?

Educated creationists acknowledge that natural selection really does occur. Creatures that do not survive do not continue to reproduce. In a competitive environment, only the best survive. Studies of gene sequences in Hawaiian fruitflies provide convincing molecular evidence in addition to the many morphological studies such as the peppered moth, Galapagos finch beaks, and several more. Natural selection in these cases prevents the degradation of the gene pool by eliminating the unfit (although the definition of unfit changes over time as survival pressures change with the environment). Therefore, natural selection is actually a conservative force to limit extremes as is demonstrated when highly bred varieties return to the wild.

Natural selection actually conserves genetic potential, when needed, so that populations can shift back to earlier expressions of the genotype (the phenotype). The important question is whether natural selection can also improve the genetic stature of a genotype. Around 1940, the respected zoologist Pierre Grassé made this comment on the creative power of natural selection (Johnson 1993):

> In spite of the intense pressure generated by artificial selection ... over whole millennia, no new species are born. A comparative study of sera, hemoglobins, blood proteins, interfertility, etc., proves that the strains remain with the same specific definition. This is not a matter of opinion or subjective classification, but a measurable reality. The fact is that selection gives tangible form to and gathers together all the varieties a genome is capable of producing, but does not constitute an innovative evolutionary process.

Evolutionary scientists often argue that given enough time, natural selection would produce the changes needed. Humans have been breeding dogs for thousands of years without producing new species, but does the acceleration of artificial selection compare with hundreds of millions of years of natural selection? Evidence suggests that brief artificial selection produces more change than millions of years of natural selection. Grassé writes, "The fruitfly ... seems not to have changed since the remotest of times," but large variations in fruitflies have been created by geneticists using artificial selection. Time cannot make up for the lack of genes; there is not a scientist in the world that can demonstrate otherwise with real evidence.

Biochemist Michael Behe of Lehigh University, in his book, *Darwin's Black Box*, writes convincingly that biological structures that are irreducibly complex cannot be evolved in a step-by-step process. An irreducibly complex object is one composed of many small parts, all of which are essential to the object's function. Behe uses the simple analogy of a mousetrap; remove any one component, even as small as the staple, and the remaining parts become useless junk. Behe then moves on to biology, giving several examples of irreducibly complex structures that could not possibly function, giving advantage in survivability, until the entire set of parts were evolved and working as an integrated whole. Behe demonstrates that natural selection working on small changes cannot produce new species.

Mathematicians have known the limitations of natural selection for a long time. As reported in *Mathematical Challenges to the Neo-Darwinian Interpretation of Evolution*, leading mathematicians at the Wistar Institute symposium in 1967 calculated the probability of a continuum of small beneficial mutations producing enough change to cross the species barrier. The results indicate the process is so improbable that it may be essentially considered to be impossible. Biological evolutionists were shocked and embarrassed to learn the mathematicians' findings. The modern dean of evolutionists, Ernst Mayr stated, "Somehow or other by adjusting these figures we will come out all right. We are comforted by the fact that evolution has occurred" (Johnson 1993).

Darwin proposed that the innate variability of life forms was sufficient to make small changes that would be locked in and propagated throughout the population by natural selection. However, later evolutionists realized that these normal variations would never create changes beyond the species barrier and that mutations were needed. Mutations, as we understand the word today, were not understood in Darwin's day. The reconciliation of pure Darwinism with modern genetic theory is known as the evolutionary synthesis, the synthetic theory, or neo-Darwinism. This modern synthesis breathed new life into evolutionary science, which was beginning to lose importance in the first half of the twentieth century. But scientists were still faced with the same overall question: can mutations followed by selection produce a continuum of development? Many think not.

In 1979, Ho and Saunders questioned the efficacy of neo-Darwinism (Behe 1996)[15]:

> It is now approximately half a century since the neo-Darwinism synthesis was formulated. A great deal of research has been carried on within the paradigm it defines. Yet the successes of the theory are limited to the minutiae of evolution,

[15]The reader may wonder why some of the honest quotes from evolutionists are somewhat dated. Although there have been advances in evolutionary science over the last 30 years, especially in molecular biology, there has been no fundamentally new understanding on how evolution works. Indeed, evolutionary science has advanced so little in its foundational principles that Origin itself remains one of the best books on the

> such as the adaptive change in coloration of moths; while it has remarkably little to say on the questions which interest us most, such as how there came to be moths in the first place.

Darwin and the modern Darwinists require that jumps across the gaps of the biological types be performed in numerous tiny steps. Therefore, the mutations of the modern synthesis would have to be very small. Darwin wrote: "Natural selection can act only by the preservation and accumulation of infinitesimally small inherited modifications, each profitable to the preserved being ...." Given the difficulty of jumping large gaps in tiny leaps, as shown by Behe, Ho and Saunders, and the mathematicians, one would think that evolutionists would abandon the "insensibly fine gradations" and go with a theory that at least had a conceptual basis: large gaps.

One highly respected (at least originally) scientist in the late 1930's and early 1940's recognized the limitations of natural selection and broke with the small steps theory. Professor Richard Goldschmidt of the University of California at Berkeley stated that many of the organs and structures we see in life today could not have arisen by the accumulation of small mutations. Although he recognized that mutations in almost every case are harmful, and large ones are utterly lethal, Goldschmidt stated that large, beneficial mutations must have occurred from time to time that bounded life forms across the species barrier. This theory has been identified as the "hopeful monster" theory. (One wonders with what the hopeful monster would mate to preserve the change.)

Goldschmidt was subjected to intense ridicule and even hate, because he was suggesting something akin to a miracle. It sounded strangely like special creation.

With amazing prescience, Darwin said years earlier, "If I were convinced that I required such additions to the theory of natural selection, I would reject it as rubbish ... I would give nothing for the theory of natural selection, if it requires miraculous additions at any one stage of descent."

After Goldschmidt, one could imagine that very few scientists were willing to depart from the fold in favor of macromutations. It would take a very strong personality indeed to withstand the storm of ridicule, but such a courageous person can be found in the outlaw of evolution, popular writer and Harvard paleontologist, Stephen Jay Gould. In 1979, Gould and Niles Eldredge recognized the lack of transitional forms in the fossil record and proposed the Theory of Punctuated Equilibrium, in which life remains in a condition of stasis most of the time but is punctuated, from time to time, by periods of rapid change. Gould and the traditional Darwinists have been sparing in the professional literature ever since. Recognizing the similarities with the despised Goldschmidt, detractors refer to Gould as Gouldschmidt.

Whether pure Darwinism, hopeful monster, or punctuated equilibrium, evolutionary science has not yet produced a coherent explanation of how the diversity of life has developed. Each is a rather imaginative scenario fraught with insuperable problems. Indeed, given the lack of supporting evidence, the lack of theoretical basis, and the imaginative nature of the theories, one could conclude that, after 150 years of development, biological evolution has as much scientific validity as special creation, in spite of the scientific venue within which evolution is discussed.

Conclusion

School children and other nonscientist members of the public (including engineers and non-biologist scientists) are constantly bombarded with the "fact of evolution." Although this chapter has not been sufficiently rigorous to refute the theory (after all, it is a theory strongly held by a large number of very bright biologists who ought to know), it may be sufficient to reveal that the evidence for biological evolution is not as coercive as public television and the biology teacher say it is. Perhaps some of the façade has been wiped away allowing individuals the opportunity for critical and more thorough evaluation of the facts.

Creationists do not have all the answers on life's origins either. For example, what did God really create? It is estimated that there are 20 to 30 million species extant in the world today, not counting the untold millions of species that have become extinct. Did God create all these individual species or did He create a few hundreds of basic genotypes that have expressed themselves in a multitude of phenotypes over time? Noah's Ark did not contain millions of species. What was the mechanism that created the diversity of life we see today and in the fossil record?

Subsequent chapters will address but probably not answer some of these questions. The next chapter speaks to the most fundamental of all biological questions, the origin of life.

Three

Origin of Life

The great physicist, Erwin Schrödinger, the namesake of the famous Schrödinger equation of quantum mechanics, once questioned why our bodies had to be so large, compared to atoms. Only a great mind would even think to ask such a question. But there is no reason why our bodies must be so large. Certainly, humans could have been made fully functional at the size of the flea on your dog. With the possible exception of the human brain, all our organs could have been made the size of a few cells, provided the remainder of the body was proportionally small.

Humans would experience a much different life in the Lilliputian world. Besides having to contend with mountainous grains of sand everywhere, we could be subject to thermal vibrations, if we were small enough. In 1827, Robert Brown observed that pollen grains in water seemed to have a means of locomotion because they were continually vibrating wildly. This Brownian motion was later determined to be the thermal energy of atoms and molecules vibrating and colliding with the pollen. In fact, in the small universe, life would be extremely violent, with individual molecules of water colliding a thousand million million times a second, seeming to shake everything to pieces.

Not only would violence reign, the really small objects, such as electrons, would appear to not have a definite location, being smeared in a probabilistic sense across space. The electrons would also appear to defy our normal understanding of physics, seeking out quantum energy levels without any apparent reason. This world would be more different than our world today than any described in science fiction books.

Yet, it is at the level of the cell and smaller, that the real story of life is portrayed. The current scientific understanding of the origin of life and the processes that make life possible are understandable only at the level of atoms, of molecules, and to some extent, of cells. Only in the last 25 years have scientists understood what was going on in the cells at the molecular level. Therefore, this chapter begins with a biochemistry primer before exploring the current evolutionary theories on the origin of life.

Biochemistry Primer

There are three major classes of the chemicals of life: proteins, carbohydrates, and lipids (fats and oils). Although life would not exist without the proper balance of all its biochemical constituents, proteins certainly play a major role. Recognizing this, the chemists of the nineteenth century named this class of chemicals from the Greek word, *proteios*, which means, "holding first place." About half of the body's dry weight is composed of proteins.

Proteins are species specific. That is, the proteins of a monkey are not the same as those of a giraffe. They are also organ specific; liver proteins differ from muscle proteins in any given individual organism. Some proteins are structural and are used like cables or building blocks. Other proteins constitute a class of biochemical catalysts known as enzymes. Enzymes enable an organism to manufacture the chemical substances necessary for life. Hormones, the chemical messengers of the body, are also made of proteins. But by far, the most fascinating function of proteins is to serve as nanomachines. In the cell, proteins are used as pumps, valves, winches, motors, information retrieval devices, power generators, sensors, and communication relays.

Proteins are very large molecules formed by chaining amino acids together into what chemists call polymers. Although there are more than a 100 amino acids found in nature, only 20 are commonly found in most proteins. Each amino acid contains about 10 to 20 atoms, with a total molecular weight averaging approximately 100. Usually, 100 to 500 amino acids are linked to make a protein, the number and order of the amino acids determining the identity of the protein. Once formed, the long amino acid chain folds upon itself into a tangled knot in which pieces of the chain weakly bond to other pieces in a minimum energy configuration. The human body manufactures about 60,000 different proteins.

When a cell "decides" it needs a particular protein, how does it determine what protein to make? After all, a protein of 100 amino acids can be arranged in far more than 10^{100} different combinations. How does a cell precisely make human hemoglobin instead of rat insulin? The marvelous answer is that each and every cell of the body contains a master blueprint that determines the characteristics of the specific individual. This blueprint, the DNA molecule, is the repository for protein manufacturing instructions, among other things.

Some time in the 1940s, scientists began to understand that the mysterious stuff in the cell's nucleus, known as deoxyribonucleic acid, somehow controls heredity. However, it took until the 1950s for Watson and Crick to elucidate the structure of DNA. This discovery eventually led to understanding that DNA segments, known as genes, contain the genetic information essential for protein synthesis.

In humans, the genetic information is encoded in 46 different DNA molecules known as chromosomes. Each DNA molecule would be about 1.5 inches long if laid out straight. However, for 46 of them to fit into a cell nucleus 10^{-4} inches in diameter, each DNA molecule must be tightly coiled on a spool. The spooled DNA resembles a long telephone handset cord. The spooled DNA is then hung like a firehose in coils. When the cell divides to reproduce, these coils of DNA are further folded onto themselves until they become the visible (under a microscope) chromosomes often seen in textbook photos. The bundling of the DNA keeps it from getting tangled during cell division, allowing the cell to distribute the DNA to the daughter cells. Once the cell has divided, the DNA unpacks from the bundles.

A DNA molecule is shaped like a fire ladder that has been twisted into a spiral or helix. The legs of the ladder are composed of alternating sugar and phosphate molecules. Each rung of the ladder is made of one of four nitrogen-containing base pairs. The bases are guanine (G), cytosine (C), adenine (A), and thymine (T). The bases are always paired across from one leg to the other in the A-T or G-C sequence.

Human DNA contains about 65 million rungs or base pairs. The order of the bases in DNA provides encoded information that specifies the amino acid sequence in protein. Every three bases identifies a specific amino acid. For example, the base sequence AAA codes for the amino acid lysine. The sequence AGT codes for serine. There is even a code to delineate the start and stop for the message: ATG means start, and TGA means stop.

DNA, the master blueprint, always stays in the cell nucleus; proteins are manufactured outside the nucleus. So how is the genetic code read and translated into amino acid sequences? The answer to this question is one of the most impressive stories of biochemistry. In a factory, the master blueprints are kept in the engineering department, but copies, the shop prints, are carried to the factory floor. In the protein factory, the DNA transcribes itself into another nucleic acid, RNA (ribonucleic acid). The so-called messenger RNA then exits the nucleus to travel to the protein-making machine.

RNA is a single-stranded molecule of very similar structure to DNA. It also has four bases except that thymine is replaced by uracil. Uracil is chemically similar to thymine and is "cheaper" to make (requires less energy).

RNA is made from DNA in a process known as transcription. First, the section of DNA containing the needed code is located, unwound, and unzipped to produce two separate strands. Special machinery made of proteins then senses the base sequence of the unzipped DNA and assembles an RNA molecule piece-by-piece except that the uracil replaces thymine. The DNA is then reassembled and the RNA is released to carry its message outside the nucleus.

The cell machine that makes protein is called a ribosome, which is a structure made of proteins and another type of RNA known as ribosomal RNA. When the messenger RNA arrives at the ribosome, it threads itself into a slot in the ribosome. The translation of the RNA code into a protein is carried out as the messenger RNA passes through a sensing device that reads the code. The amino acids in the protein are then assembled together having been brought to the ribosome by an army of specially fitted transfer RNA molecules. When the codeword, stop, is encountered, the completed protein is released and then folds itself into its customary shape. The messenger RNA is then released, allowing the ribosome to clear itself for manufacture of another protein. Most proteins can be manufactured in this way in about three minutes. It is difficult to reconcile such a well-designed and intricate process with random atomic interactions.

There is yet one more fascinating story to tell about this cellular, biochemical world. A cell contains only one master blueprint. Thus, when the cell divides, another copy of DNA must be produced in a process called replication. If the copy is not perfect, the organism degrades or dies.

Just as in transcription, for DNA to replicate, it must unwind and unzip. Each of the two strands is then fed through a copying mechanism which reads the bases and attaches the complementary base, sugar, and phosphate molecule. The copying mechanism contains "proofreading" proteins that ensure there are no base mismatches or damaged parts. Each of the two DNA molecules consists of one new strand and one original strand. Following replication, the cell divides with identical DNA ending up in the nucleus of each cell.

No matter how diverse or bizarre the form, all life has the same methods of replication, transcription, and translation - the same methodology in the genetic code. This condition is strongly suggestive of either a common design feature or of the uniqueness of the origin of life[16]. If life developed from non-life more than once, it would be strange indeed if the amazingly intricate cytology and genetic mechanisms would be identical in every branch from original life. Furthermore, from the viewpoint of the evolutionist, the large differences between the animal and plant kingdoms are suggestive of two different origins. Yet, their identical genetic mechanisms are suggestive of common origin. How much more elegant and straightforward to understand life in the context of an intelligent Designer and Creator!

The Oparin Hypothesis

Recognizing that the gap between non-life and life is orders of magnitude greater than the gaps between the major taxonomic categories, Darwin said, "It is mere

[16] Biologists believe that life originated once and diversified from the original life form. This is because of the isomeric properties of many organic molecules. Isomerism is discussed in the next section.

rubbish thinking at present of the origin of life; one might as well think of the origin of matter" (Britannica 2000). Darwin's statement does not mean that he did not believe in a mechanistic origin of life, but that such considerations were far too advanced for science of his day. Nevertheless, there was a prevailing view among scientists that life spontaneously developed from non-life, under the right circumstances. It was Louis Pasteur who, in 1864, finally demonstrated that spontaneous generation was not a valid scientific explanation for certain observed biological phenomena.

Nevertheless, the idea of spontaneous generation has never died and even flourishes today, albeit in a different form. In 1871, Darwin himself speculated that life might develop from non-life (F. Darwin, 1887):

> It is often said that all the conditions for the first production of a living organism are now present, which could ever have been present. But if (and oh! What a big if!) we could conceive in some warm little pond, with all sorts of ammonia and phosphoric salts, light, heat, electricity, etc., present, that a protein compound was chemically formed ready to undergo still more complex changes, at the present day such matter would be instantly devoured or absorbed, which would not have been the case before living creatures were formed.

Fifty years later, Russian biochemist Aleksandr Ivanovich Oparin presented a hypothesis for the origin of life that resembled Darwin's speculation. He reasoned that the primitive Earth's atmosphere must have consisted of large amounts of ammonia, methane, hydrogen, water vapor, but little or no oxygen. Chemists would call this a chemically reducing environment[17]. Given some lightening, heat, and ultraviolet light to overcome the activation energy, some complex organic molecules might be spontaneously produced. Oparin continued to reason that, over time, these complex molecules would react to eventually form the proverbial prebiotic soup. As the organic molecules aggregated, they eventually assumed cell-like behavior and life began. British chemist J. B. S. Haldane proposed similar ideas about the same time.

Oparin's hypothesis, which was based on a combination of speculation and knowledge of biochemistry, was warmly received by chemists and biologists alike. Its analogy to and continuity with Darwin's biological evolution produced a very intellectually satisfying proposal. However, it wasn't until the early 1950s that anyone attempted to experimentally substantiate Oparin's hypothesis. In 1950, graduate student Stanley Miller at the University of Chicago was listening to a lecture by Nobel laureate Harold Urey in which Urey alluded to the Oparin hypothesis. Miller decided to test the hypothesis with Urey's guidance.

Miller set up an apparatus with Oparin's formula for the supposed early earth atmosphere and sparked the atmosphere to simulate lightning. Miller was able to extract amino acids and other organic molecules of

[17] A reducing environment is one in which oxygen is not present to react with the chemicals present. For example, the presence of iron sulfide, FeS2, in the Earth's crust is in a reduced state and indicates that the molecule formed in the absence of oxygen. Since hydrogen is an oxygen scavenger, an atmosphere with large amounts of hydrogen would be a reducing environment. The opposite condition, an oxidizing environment, is one in which chemicals tend to react with oxygen. In our example, we find a great abundance in the Earth's crust of iron oxide, Fe2O3, which is one of iron's oxidized states.

interest[18]. Miller's experiment, which was published in *Science* in 1953, resulted in a number of careers dedicated to similar experiments. Over the years, a large body of data has been collected on organic synthesis in what is believed to be prebiotic conditions. Evolutionists look to Miller's experiment as a milestone in the quest for understanding the origin of life. Therefore, it is appropriate to understand just what Miller's experiment, and others since then, have contributed to mankind's understanding of the origin of life. Does it establish the Oparin hypothesis?

In performing the genre of experiments like the Miller experiment, scientists have been gathering data on a hypothesis. This is good application of the scientific method. However, the area of potential difficulty is the rushing to judgment on whether the Oparin hypothesis has been tested and found to be substantiated. If the Oparin hypothesis were true, scientists should eventually find geochemical evidence of the ancient reducing atmosphere. Although they may encounter difficulties, and it may take many years, eventually scientists should be able to adequately mimic the presumed constituents of the early organic soup and produce important cellular building blocks. Collecting this kind of evidence builds support for the hypothesis, but does not validate it until the body of evidence is overwhelming. On the other hand, getting some few results consistent with Oparin's hypothesis in the face of insuperable difficulties with the hypothesis clearly does not substantiate it and may tend to invalidate it.

---

[18] Miller's first few attempts did not produce any organics of interest. He adjusted the experiment until it produced amino acids.

Difficulties with the Hypothesis

Many of the difficulties scientists have found with Oparin's hypothesis are technically beyond the scope of this discussion. Nevertheless, six of the more simple challenges are described below. Although origin-of-life scientists could conceivably work their way around these difficulties, the problems with the hypothesis are great and should provide impetus for honest scientists and honest producers of public education programs to cease presenting the prebiotic soup idea as proven fact.

1. Scientists have long speculated that the original Earth's atmosphere was more like that of the Jovian planets, containing hydrogen, methane, and ammonia[19]. This would mean that in the reducing atmosphere, carbon would not exist as carbon dioxide (CO_2) but as methane (CH_4); nitrogen would not exist as molecular nitrogen (N_2) but as NH_3 (ammonia). Any oxygen (O_2) would exist as water (H_2O). The oceans of the world would also be largely composed of chemicals in their reduced state. If the Oparin hypothesis is true, then the organic soup in the oceans would have to exist in concentrated form for many millions of years. As sediments were laid down, molecules of the reducing organic soup would have been trapped within the sediments. However, geochemists have not been able to find evidence of such organic molecules in the oldest sediments where they should be. Also, the great abundance of minerals in the crust are in oxidized forms. This one weakness in the Oparin hypothesis should be enough to raise serious doubts about its validity.

[19] The Earth's lesser gravity and warmth, due to nearness to the Sun, is presumed to have caused the hydrogen and other light gaseous substances to escape into space. This loss of hydrogen, along with oxygen from the early photosynthetic organisms, would have transformed the atmosphere into the current oxidizing atmosphere.

2. As we have seen, the hypothesis requires the atmosphere to contain no oxygen. The organic chemicals needed simply will not form in the presence of oxygen. (The reaction in reducing conditions is exothermic; the reaction in oxidizing conditions is endothermic.) However, in the absence of free oxygen, the atmosphere would not contain ozone (O_3). Without the ozone layer, the Earth would have been bathed in lethal ultraviolet light. Not only would life be sterilized, the organic products would be broken down as quickly as they were formed. This produces a difficult situation for the hypothesis. If there were oxygen in the atmosphere, the organics could not form; if there were no oxygen in the atmosphere, the organics would break down. The hypothesis doesn't work either way. In fact, all Miller-style experiments are performed with a trap to quickly remove any organics formed so that the products won't degrade under the influence of the energy causing the reactions to proceed.

3. Asymmetric organic molecules often form in two different arrangements in mirror images. These different forms are known as isomers. Chemists refer to the two forms as left-handed and right-handed isomers. For example, the essential biochemical, glucose, comes in a left-handed form (levulose) and a right-handed form (dextrose). However, the body only uses dextrose. Similarly, amino acids come in left-handed and right-handed varieties. Miller-style experiments produce both types of amino acids. Strangely, only left-handed amino acids are used in biologically important proteins. How would early proto-life manage to select only the left-handed amino acids for proteins? In the construction of a protein containing hundreds of amino acids, how does random synthesis consistently select only one variety?

4. Another similar difficulty as with the isomers, is the chemical bonding between the amino acids. Amino acids can bond to each other in several ways. However, in biologically important proteins, only the peptide bond is used. Again, one should ask, how do random chemical reactions produce a polymer hundreds of units long with the same bond used consistently every time? We have seen that the amino acids in a protein also weakly bond to each other from the side (in addition to strong bonds end-to-end) to fold the chain into a knot. In an organic soup, these weak side bonds would be filled with other constituents in the soup rather than other parts of the protein molecule. How is it that biologically important proteins formed within a soup are able to avoid bonding with other species and fold into the biologically important structures?[20]

5. Studies on the minimum size and complexity of a cell indicate that what is needed to produce a self-replicating entity is far too complex for a chance agglomeration of chemicals to produce. The minimum cell contains intricate structures that just don't evolve. Even if one could artificially manufacture a crude, self-replicating proto-cell, its reproductive machinery would produce large numbers of translation errors. Because hundreds of proteins would have to work in concert with varying functions, poorly made protein would not work. If it managed to reproduce, each generation would be worse than

---

[20] Many biochemists recognizing this difficulty, have proposed that the RNA must have been synthesized first, then the RNA was used to translate into the protein. Origin-of-life biochemists are largely divided into two camps: the protein-first and the RNA-first camps. There are insuperable difficulties either way. Synthesis of RNA requires the construction of several large molecules that comprise parts of RNA; however, the conditions needed for their synthesis are incompatible with each other.

the previous one. The entire system would fail. In the words of evolutionist Carl Woese, "The primitive cell was faced with the seeming paradox that in order to develop a more accurate translational apparatus, it had first to translate more accurately" (Denton 1986). Complex systems just do not gradually evolve in step-by-step fashion because of the necessity for integrated operation of the many parts through every phase of the evolution.

6. The joining of amino acids into the protein polymer results in a waste molecule of water. Thus, water actually inhibits the polymerization of amino acids. Of course, this poses difficulties for the organic soup hypothesis. Therefore, scientist Sidney Fox has speculated that the organic soup splashed on the hot rim of a volcano, evaporating the water as quickly as it was formed in the polymerization process. This improbable scenario has the difficulty that the heat would also destroy the process needed to make the protein. The Oparin hypothesis needs the water to form the soup but the very existence of water inhibits the reaction.

Scientists are well aware of the difficulties in origin-of-life theories that are presented here. The difficulties are so great that some Nobel-prize-winning scientists (Francis Crick of DNA fame and astronomer Fred Hoyle are two examples) have given up on the Oparin hypothesis and have been propagating the idea of panspermia. This concept states that life on Earth was seeded from outer space, either by aliens or by a meteorite. Of course, panspermia does not solve the problem, but merely shifts it to another place and time. That world-class scientists would resort to such an idea is an indication of how badly scientists want to believe in a theory that cannot be substantiated.

Molecular biologist Michael Denton, in his book, *Evolution: A Theory in Crisis*, compiles these quotes from evolutionary scientists about the difficulties of origin-of-life theories.

> The development of the metabolic system which as the primordial soup thinned must have "learned" to mobilize chemical potential and to synthesize the cellular components poses Herculean problems. So does the emergence of a selectively permeable membrane without which there can be no viable cell. But the major problem is the origin of the genetic code and of its translational mechanism. Indeed it is not so much a problem as a veritable enigma. [J. Monod in *Chance and Necessity*]

> An honest man, armed with all the knowledge available to us now, could only state that in some sense, the origin of life appears at the moment to be almost a miracle, so many are the conditions which would have had to have been satisfied to get it going. [Francis Crick in *Life Itself*]

As we learned in Chapter 2, there is a gap between the great types of nature that cannot be spanned through evolutionary processes. The quotes above indicate that scientists understand that the gaps between the phyla or even the biological kingdoms are miniscule compared to the gap between life and non-life. Recent discoveries in molecular biology only serve to highlight the magnitude of the gap.

Even though there is order and beauty in a single snowflake, the simplest of living organisms, a bacterium, is as much more complex than a snowflake than an entire city is to a pile of lumber. Crystalized water provides no comparison in complexity. Furthermore, no cells known today can be considered primitive. Although there are important differences, all are equally complex - from those of a bacterium to those of a human. There are no precursors. All cells undergo transcription, translation, and replication. How so many scientists propagate the "fact" of prebiotic evolution remains a mystery.

A Survey of the Literature

Biochemist Michael Behe of Lehigh University is one of the insiders that have courageously "come out of the closet" regarding the limitations of Darwinian theory and molecular evolution. His book, *Darwin's Black Box, the Biochemical Challenge to Evolution*, provides a very lively and creative discussion on many of the issues discussed in this chapter. In addition, he provides a survey of the literature of molecular evolution that is illuminating.

The *Journal of Molecular Evolution* was established in 1971. Over the last 10 years, it has published approximately 1,000 papers on molecular evolution. These papers roughly categorize into three types: (1) chemical synthesis of molecules thought to be important to the origin of life, (2) abstract mathematical models, and (3) comparison of DNA or protein sequences. Behe then discusses the issues and difficulties with each category. For category 1, he briefly mentions some of the issues discussed in the subsection above and quotes respected biochemist Klaus Dose:

> More than 30 years of experimentation on the origin of life in the files of chemical and molecular evolution have led to a better perception of the immensity of the problem of the origin of life on Earth rather than its solution. At present, all discussions on principal theories and experiments in the field either end in stalemate or in confession of ignorance. ["The Origin of Life: More Questions than Answers," *Interdisciplinary Science Reviews*]

Behe explains that the mathematical models of category 2 are useful for exploring what might occur under presumed conditions, but can tell the investigator nothing about what really happened. Finally, the comparison of DNA and protein sequences are interesting but, as discussed in Chapter 2, reveal a typological nature of life at the biochemical level (although Behe allows that common descent could be indicated).

Although much of what was published is good science, the bare fact is that none of the articles has any explanation on how complex structures developed through mutation and natural selection. Not one paper proposes a model by which complex biochemical systems evolved.

Behe then goes to the *Proceedings of the National Academy of Sciences*, which published 20,000 papers between 1984 and 1994. About 400 of these papers were on molecular evolution; none of them detail routes for the origin of complex biochemical structures. Next, Behe discusses books. Again, none provide any mechanisms by which real cellular systems could have originated. This plethora of publishing, in the absence of any real progress or results, is, in Behe's colorful style, "hope struggling valiantly against experimental data."

Conclusion

Darwinian evolution is a mechanistic philosophy for the origin of life's diversity. As such, it is incompatible with any miraculous theory for the origin of life. If the origin of life cannot be explained in mechanistic terms, then Darwinian evolution must limp along, as a man on one leg, until a prosthesis is supplied.

We have seen that origin-of-life science is faced with difficulties even greater than those faced by Darwinian evolution. Perhaps, one day scientists will solve these difficulties, but many evolutionary scientists today discuss the origin of life in terms like "miracle." It is no wonder that respected insider scientists like Michael Behe and Michael Denton are beginning to challenge the foundations of evolutionary science. Because of the monumental challenges to evolution, scientists at university campuses and research centers are beginning to discuss the concept of intelligent design, the topic for the next chapter. These scientists are very much in a minority, but cracks in the evolutionary edifice are beginning to show.

Is it possible that one day scientists will not be threatened by belief in a Creator God, believing that deep and fulfilling inquiry into science is not thus limited? God must marvel at the arrogance of mechanistic scientists as He did with Job: "*Where wast thou when I laid the foundations of the earth? Declare, if thou has understanding*" Job 38:4.

Four

Paley and Dawkins Again

The Argument for Intelligent Design

It was in 1959 that the visionary and colorful physicist, Richard Feynman of Cal Tech, issued a challenge to his colleagues of the American Physical Society (Regis 1995):

> It is my intention to offer a prize to the first guy who can take the information on the page of a book and put it on an area 1/25,000 smaller in linear scale in such a manner that it can be read by an electron microscope.
>
> And I want to offer another prize ... to the first guy who makes an operating electric

> motor - a rotating electric motor which can be controlled from the outside and, not counting the lead-in wires, is only 1/64 inch cube.

A motor the size of a period at the end of a sentence was conceptually possible, but it was not clear whether it could be manufactured. The 1/25,000 scale book page would be small enough for the entire *Encyclopedia Britannica* to fit onto the head of a pin. The fulfillment of that challenge seemed very far off indeed. However, the next year, Bill McLellan walked into Feynman's office and collected his thousand dollars for the motor. It took until 1985, but Tom Newman collected the second prize by writing the first page of *A Tale of Two Cities* so small that individual letters were only about 50 atoms across.

As impressive as these feats are, they are only examples of *micro*technology. Scientists today are pushing the limits even further with *nano*technology. In 1990, scientists at IBM spelled out the IBM logo using individual atoms - 35 of them. Later they created a map of the Western Hemisphere at a 1:10,000,000,000,000 (one to 10 trillion) scale. But these are only gimmicks. Scientists are envisioning the creation of nanomachines that would change human life more than all the technology of the past combined.

Nanotechnology is the design and manufacture of hypothetical molecular-level machines, including so-called assemblers, that move individual atoms around to suit, building other nanomachines, including copies of themselves. With nanotechnology, every individual on the face of the Earth could afford to make anything he or she wanted, basically free - just like on the starship, *Enterprise* with the fictional "replicators". Nanomachines could become permanent residents of the body, repairing the effects of aging and disease as they occur, giving humanity eternal life. The cryogenically-suspended dead could be thawed and the freeze damage repaired by the nanomachines. (There are already frozen people unconsciously awaiting this hopeful scenario.) Any item of food could be produced in minutes, eliminating hunger with meals of perfect taste and texture.

These prospects might seem remote, but there is a considerable body of scientists working on nanotechnology right now. Nanotechnology has risen from a fringe science with almost religious overtones to a respectable science. The difficulty is that scientists have not yet figured out a way to design and construct a nanomachine. There are conceptual designs on the drawing board, but no designs that could be taken to the "shop" for construction. One hopeful avenue is to design proteins that would fold into appropriate shape that could serve as machines or parts of machines. However, the design problem of getting proteins to cooperate are immense - hence no nanomachines yet.

Working on nanotechnology is commendable, but our scientists are too late. We already have wonderfully complex and efficient nanomachines working miracles of technology - in every cell of every living entity. Since the discovery of the structure of DNA in 1953, molecular biologists have been discovering that life is based on the operation of nanomachines. Biochemist Michael Behe tells us (Behe 1996):

> Molecular machines haul cargo from one place in the cell to another along "highways" made of other molecules, while still others act as cables, ropes, and pulleys to hold the cell in shape. Machines turn cellular switches on and off, sometimes killing the cell or causing it to grow.

> Solar-powered machines capture the energy of photons and store it in chemicals. Electrical machines allow current to flow through nerves. Manufacturing machines build other molecular machines, as well as themselves. Cells swim using machines, copy themselves with machinery, ingest food with machinery. In short, highly sophisticated molecular machines control every cellular process. Thus the details of life are finely calibrated, and the machinery of life enormously complex.

The details of DNA replication and transcription into RNA and the translation of genetic instructions into proteins (see Chapter 3) reveal remarkable nanotechnology that is compelling evidence for design. However, some evolutionists such as Richard Dawkins are not impressed that engineering found in living systems is evidence of design. As we learned in Chapter 1, Dawkins says (Dawkins 1996), "Biology is the study of complicated things that give the appearance of having been designed for a purpose." So how does one recognize design?

Recognizing Design

How does Dawkins look at life with "the appearance of having been designed" and conclude random processes were at work and Paley, Behe, and others examine the same information and conclude intelligent design? If something appears to be designed, then design should remain a working hypothesis until further study has coercively ruled it out. However, the science establishment today makes the *a priori* decision to eliminate design as a potential cause. We have seen in Chapters 1 through 3 that there is nothing that compels one to abandon intelligent design - that descent with modification through random causes is really a philosophical presupposition.

Again, how does one distinguish design from random occurrence? A snowflake is a beautiful six-sided crystal that moves many people to conclude that a Designer was involved. We were told in kindergarten that no two snowflakes are alike. Although this maxim may not be true, it leads one to further marvel at the miracle of creation. However, does a snowflake really compel an objective observer to conclude that the snowflake was designed? It is, after all, the natural result of well known physical laws, working as expected. For those with a religious bias, snowflakes are evidence of the Creator's hand, but should we expect Richard Dawkins and his colleagues with purely mechanistic bias to be impressed? Probably not.

Another example, which William Paley used repeatedly, is the anatomy of the eye. Using eighteenth century knowledge of the invertebrate eye, Paley built a compelling case for the design of the eye, likening it to the design of a telescope. Dawkins, knowing so much more of its anatomy and physiology, is even more impressed with the eye. "Here Paley could have gone even further," he writes about the apparent design of complex biological systems. Dawkins then launches into a description of the eye that inspires the reader even more than Paley's does.[21] However, as marvelous as is the eye's design, Dawkins concludes, "The analogy between telescope and eye, between watch and living

---

[21] As a zoologist, even Dawkins fails to see the real story of the eye that can only be told at the nano level. The molecular transformation of light energy to electrical energy in numerous complex steps, as told by Michael Behe, is even more impressive.

organ is, false." One must ask if there is anything in biology that would cause a committed evolutionist to see design.

Creationists[22] have long cited macro examples such as the eye, the bombardier beetle, the life cycles of creatures with complex symbiotic relationships, and even the existence of sexual reproduction as evidence for intelligent design. However, the evolutionists have always explained these examples by storytelling, a step-by-step imagined biological history. Such storytelling abounds in the popular evolutionary literature - especially on public television. One of the first in the genre of storytelling is from Darwin himself and is typical of what is often found today. Here is Dawkins' paraphrase of Darwin on the eye (Dawkins 1996):

> Some single-celled animals have a light-sensitive spot with a little pigment behind it. The screen shields it from light coming from one direction, which gives it some "idea" of where the light is coming from. Among many-celled animals ... the pigment-backed light-sensitive cells are set in a little cup. This gives slightly better direction-finding capability. ... Now, if you make a cup very deep and turn the sides over, you eventually make a lensless pinhole camera. ... When you have a cup for an eye, almost any vaguely con-

[22] For variety, this chapter refers to creationists and proponents of intelligent design in a seemingly interchangeable manner. However, there is a difference. Creationists believe that the universe was created by a personable God who is interested in the affairs of mankind. Among creationists, there are biblical literalists who believe in a six-day creation and others who generally agree with the prevailing cosmic and geologic theories. Proponents of intelligent design may be creationists, but are more likely to be deists, theistic evolutionists, or agnostics. Similarly, Darwinists are the majority subset of evolutionists that believe evolution largely as Darwin and now Richard Dawkins teach it.

> vex, vaguely transparent or even translucent material over its opening will constitute an improvement, because of its slight lens-like properties. Once such a crude proto-lense is there, there is a continuously graded series of improvements, thickening it and making it more transparent and less distorting, the trend culminating in what we would all recognize as a true lens.

Thus, it seems, no matter how marvelous creation appears to be, no matter how complex the organ or function, it is impossible to produce an example that will compel an evolutionist to abandon Darwinism. Darwin himself knew that any such example would falsify his theory. He wrote in *Origin*, "If it could be demonstrated that any complex organ existed which could not possibly have been formed by numerous, successive, slight modifications, my theory would absolutely break down." Yet the theory never breaks down, because of storytelling - that is, until Michael Behe.

In his 1996 book, *Darwin's Black Box*, Behe may have finally answered the question on recognizing intelligent design. Behe is not a creationist, but he believes that a Mind is behind the complexity of life.[23] Behe defines a term that describes structures that are truly evidence for intelligent design: irreducible complexity - "a single system composed of several well-matched, interacting parts that contribute to the basic function, wherein the

[23] It is too bad that Behe has not further concluded that the God who claims to be the Author of life and the author of the Bible is his Intelligent Designer. The Scriptures can explain many of the mysteries of a deistic intelligent design theory such as the imperfections of nature and the existence of evil. Later in this chapter, we will discuss imperfection. For a discussion on what goes on behind the scenes to cause imperfection and sorrow in the world, see the author's unpublished, undated essay, "Why Bad Things Happen to Good People."

removal of any one of the parts causes the system to effectively cease functioning." Such a system cannot be developed step-by-step by small, successive improvements while maintaining function as it evolves. Natural selection can only work on existing functioning systems. Therefore, the system has no way to originate as an integrated functioning entity.

Being a master of analogy, Behe uses a mousetrap as an example. The first step in determining whether a system is irreducibly complex is to specify its function and identify all the parts and their purposes. For the mousetrap, there is (1) a wooden platform to act as a base, (2) a metal hammer to kill the mouse, (3) a spring to provide the motive force, (4) a sensitive catch to hold the bait and release the hammer at the appropriate time, and (5) a metal latch to hold the hammer until the catch signals the appropriate time. All these parts work together to perform the mousetrap's function: to kill a mouse.

The second step in identifying irreducible complexity is to determine if every component is required to fulfill the system's function. In the case of the mousetrap, it is easy to conceptualize the system and recognize that each and every component is necessary for the system to function. The individual parts do not need to be optimally designed, but they all have to be sufficient to fulfil their purposes. For example, a hammer delicately carved of wood might still yield a functioning mousetrap (although it would probably only function once), but a coil spring made of wood would not suffice at all.

Now, an existing mousetrap might "evolve" into a better one by small successive changes, all the while maintaining functionality at every step. For example, our mousetrap with a wooden hammer might be modified to one with a metal hammer, but there is no functioning precursor mousetrap that could change in a single small step into the specified mousetrap. A mousetrap without a spring does not kill mice poorly; it doesn't kill at all.

The mousetrap is then an irreducibly complex system. All the components must be brought together at once and assembled in the correct arrangement to be functional. For such an occurrence to happen by chance is a miracle - something unacceptable to evolutionists. The more complex the system, the greater the miracle needed and the stronger the argument for intelligent design.

Behe then describes five examples of biological irreducible complexity: (1) the cilium, which some cells use to swim, (2) the blood clotting mechanism, (3) subcellular transport of materials, (4) parts of the immune system, and (5) the metabolic pathway for AMP synthesis. Behe selected molecular examples because the macro examples are so complex that the ambiguity previously described often results. His examples are described in Behe's book in some detail, a chapter for each one. The reader is encouraged to read further to see how impossible evolution is in comparison to intelligent design. The message of the nanomachines Behe describes is that, even more surely than Paley's watch, life was designed.

There is yet one more answer to our question on recognizing intelligent design. That answer comes from information theory. Imagine that you are walking on a deserted beach, hunting for sea shells. You come across some shells arranged perfectly in the capital letter "I." A thoughtful person would probably believe that someone had arranged the shells, but his confidence in that opinion would depend on how many shells were involved. As few as 10 shells could be a chance occurrence. With 50 shells, perhaps there would be less doubt.

Now suppose you find two letters spelled out in this order: "It." Perhaps even Richard Dawkins would conclude that someone had arranged the shells. Let's take it a step further. Suppose you find written on the sand, "It was the best of times, it was the worst of times." There is now no doubt in anyone's mind that the shells were arranged. You might even draw some conclusions about the nature of the arranger.

What made the difference in your level of confidence that intelligent design was behind this coastal artifact? Was it not the amount of information involved? A single letter raises the question. Two letters nearly answer the question. An entire sentence is tantamount to absolute proof, even though there was no scientific observation on the placement of the shells.

How much information must be in a human zygote? It is the single-celled beginning of an entire person. Where are the instructions for the person's manufacture? How does just a single cell "know" how to make kidneys and eyes and a brain? How does it know how to make the immune system with all of its interdependent parts? How does it know how to tie all the systems of a human body into a fully integrated functioning entity? The instructions must be exceedingly complex indeed! Such an immense quantity of information is contained within the tiny nucleus of that cell, in just 46 molecules.

The information encoded on our hypothetical beach by geometric symbols is trivial compared to the amount of information encoded by molecules of DNA. Whereas two or three letters on the beach are sufficient for us to exclaim that a designer was at work, what about untold billions of pages of information in DNA? The Psalmist could not have known but a fraction of the truth when he exclaimed, *"I am fearfully and wonderfully made."* Psalm 139:14.

Arguments Against Design

There are four general arguments against intelligent design; they are all philosophical - not scientific. The first is British philosopher David Hume's classic rebuttal. The second is the argument from imperfection. The third is the quasi-scientific argument that, given enough time, even improbable events will occur. Finally, evolutionists say that design is outside the bounds of science. The following subsections discuss the first three of these objections to design. The fourth argument is addressed in Chapter 8.

David Hume

David Hume (1711 - 1776) was a Scottish empiricist philosopher and skeptic. In 1779, Hume's *Dialogues Concerning Natural Religion* was posthumously published (to avoid harsh criticism) in which Paley's design argument was refuted 23 years before Paley wrote his *Natural Theology*. Hume's criticisms of intelligent design have been cited ever since as being devastating to the concept of natural theology. Among his arguments are three of interest to us. These are paraphrased below in modern terms.

Argument one: The universe, and especially Earth, seems to be optimized for life. There is a careful balance in the parameters of the universe that is critical to life, giving the appearance of, at least for the Earth, being fabricated for mankind. However, humans should not present the optimized universe as evidence for design, because if it had been any other way, there would have been no one to marvel at the optimization. When creationists argue from the viewpoint of physics being optimized for life, they are in effect offering a tautology.

This objection to design is akin to the situation on the beach in which we wonder how many letters must be spelled out before we conclude design. As described by Denton (1998), there are a huge number of physical parameters that must be perfectly balanced for life to exist. At what point do we specify a universe so uniquely that design is the only logical conclusion? For the religious person, not many specifications are necessary at all. For the skeptic, the number is infinite. Given this ambiguity, one might have to award the point to Hume; however, yielding this point is not the same as yielding the case for intelligent design.

Argument two: The particles and physical constants of the universe are in constant motion. Occasionally, a configuration or system state is reached that is metastable and remains so for a sufficiently long time for it to express itself in the outgrowth of its natural laws. Given an infinite amount of time, every conceivable metastable state would be achieved. At least one of them would be expected to produce life and indeed has.

This objection has been repeated in modern times as the parallel universe theory, in which every conceivable system state exists simultaneously. In those suitable for life, life exists. Of course there is neither empirical nor theoretical support for the multiple state concept other than the statistical mechanics of gases, which is an entirely different matter altogether. The argument that, given enough time, the impossible will happen is identical to the third general argument against design and is addressed later in this subsection. Hume makes a poor showing on this objection.

Argument three: Machines are obviously designed and manufactured entities. However, life only superficially resembles machines; it is really natural. The analogy of living things with machines only holds if life is profoundly like machines. Thus, arguments for intelligent design of living things based on complexity are not valid.

It is this objection to design theory that has plagued creationists the most and has been the subject of much of this chapter. Philosopher Paul Nelson, of the University of Chicago, recognizes the difficulty creationists face with this argument of Hume's (Nelson, 1998): "Thus design must ever be an explanation from ignorance, attempting to prove a negative and vulnerable at every instant to being overturned by fresh knowledge in science." But this defensive posture by creationists is artificial. They have allowed practitioners of naturalism to box them into a philosophical corner.

As Dr. Nelson points out, skeptics of design worry that the design hypothesis prematurely shuts off scientific inquiry about origins. However, in many cases, the skeptics do not have a naturalistic alternative to fall back on. If a natural cause could have been postulated, it would have been. Under these circumstances skeptics place an unrealistic standard of certainty on the creationist, to prove the negative that life did not result from random occurrences. In reality, intelligent design is as viable a scientific alternative as any naturalistic one. In the face of insuperable problems for causation from random occurrence, intelligent design is the one much preferred. Skeptics make an *a priori* decision to eliminate from further consideration a hypothesis that could very well be true, regardless of how distasteful the hypothesis is philosophically.

Also, Dr. Behe has shown that, at the molecular level, life really is profoundly like a machine. Hume and more recent skeptics have underestimated the power of the analogy. Even Hume would not argue that a watch was the result of random interactions of atoms. Yet, at the molecular level, we have even more profound examples of irreducible complexity that demand, at least equally strongly, that life resulted from an intelligent Maker.

The Argument from Imperfection

Intelligence sufficient to design living systems must be extraordinarily intelligent. Thus, one would expect such intelligence to produce perfect designs, as least as far as humans could judge. However, evolution, being a trial and error effort, should occasionally produce suboptimal designs. Nature shows evidence of suboptimal designs. Therefore, say the evolutionists, intelligent design is false and evolution is true. (This subject was introduced in Chapter 2.)

Because creationists have for so long appealed to teleology, evolutionists have heaped scorn[24] on intelligent design proponents by citing examples of purported poor design. Since the time of Paley, the eye has been a favorite of creationists in promoting intelligent design. Consequently, it is also a favorite of evolutionists in the argument from imperfection.

[24] Creationists should not be offended by the scorn, for creationists have often been the kings of ridicule, sometimes implicitly denigrating the Darwinists as fools. How better it would be if all sides in the debate dealt with the issues alone. In this book, I have tried to treat the evolutionists with respect. Many of the champions of evolutionary theory, such as Ernst Mayr, are honest intellectual giants. Even though I have poked some fun at that most staunch of all Darwinists, Richard Dawkins, I truly hope that Dr. Dawkins takes it gently, in the spirit in which it was offered.

Richard Dawkins writes (Dawkins 1996):

> Any engineer would naturally assume that the photocells would point toward the light, with their wires leading backwards towards the brain. He would laugh at any suggestion that the photocells might point away from the light, with their wires departing on the side nearest the light. Yet this is exactly what happens in all vertebrate retinas. Each photocell is, in effect, wired in backwards, with its wire sticking out on the side nearest the light. The wire has to travel over the surface of the retina, to a point where it dives through a hole in the retina (the so-called 'blind spot') to join the optic nerve. This means that the light, instead of being granted an unrestricted passage to the photocells, has to pass through a forest of connecting wires, presumably suffering at least some attenuation and distortion (actually probably not much but, still, it is the *principle* of the thing that would offend any tidy-minded engineer!).

Many other examples could be cited, including the panda's thumb (Gould 1980), pseudogenes, noncoding DNA, vestigial structures, and the coevolutionary "arms races" among antagonistic species (Futuyma 1998).

The argument from imperfection may seem compelling at first glance, but it has no substance. There are four basic flaws in the argument. The rebuttals come from a mixture of engineering, philosophy, and theology.

Flaw 1: The imperfection argument assumes an understanding of the Designer's desired endpoint. With amazing prescience, evolutionists seem to know God's purpose in creating life. Humans cannot read the mind of an omniscient and omnipotent God. "*For as the heavens are higher than the earth, so are my ways higher than your ways, and my thoughts than your thoughts.*" Isaiah 55:9.

Flaw 2: Mankind has never designed one living entity. Therefore, what credentials do evolutionists bring to the debate in presuming to distinguish what is a perfect creation and what is not? After all, these so-called flawed designs do function pretty well. The healthy human eye, while not as good as the eagle's in some respects, nor the fly's in other respects, is pretty well adapted to human needs[25]. Pure Darwinism places a heavy emphasis on survivability, but perhaps the design of life is not optimized for survivability. (There are theological arguments that would say definitely not!) Do humans even know what perfection is? Would a perfect creature never die? Produce infinite offspring? Never produce waste products? Hear all sound frequencies and see the entire electromagnetic spectrum? It is clearly arrogant to judge the Creator's works as flawed.

Flaw 3: If we grant the point that some life designs are flawed, even a poorly designed system has a designer. Darwinism has not provided any alternative mechanism for the development of irreducibly complex systems.

Flaw 4: Although evolutionists will not be impressed with this flaw, it is offered nonetheless. From a theological viewpoint, there is a good reason for the flaws and suffering present in this world. A hint is found in Genesis 3, Revelation 12, Job 1 and other scriptural ref-

[25] Peter Gurney offers some good explanations for the eye's design in, "Is Our 'Inverted' Retina Really 'Bad Design'?" in *Creation Ex Nihilo Technical Journal*, Volume 13, Number 1.

erences (Connor, unpublished essay available on request).

Overcoming Improbability

It is true that given a sufficient number of trials, improbable events can occur. If one performed 1,000 repetitions of a 10-trial coin flip (10,000 flips), one might just begin to get a reasonable probability of heads 10 times in a row. Evolutionists acknowledge that the probabilities in the origin of life and its evolution are exceedingly small, possibly vanishingly small, but because we exist and we had to have evolved by chance, evolution must be possible. They say, given enough time, evolution can occur.

Since we cannot intellectually grasp probabilities as low as we are talking about, it is time for an analogy. Let's pick on Dickens yet a third time. Imagine a high school book report on *The Tale of Two Cities*. We type it into an especially fast computer that is programmed to evolve the report into Dickens' actual text. The algorithm attempts to change one word at a time, randomly, either by addition, deletion, or substitution. After each random attempt, the program examines the result and decides if the change is an improvement towards the goal. If it is, then the change is accepted and the next change is attempted. If it is not, the change is rejected.

How long do you think it will take before the book report evolves into the novel? The computer is very fast so thousands of changes can be made per minute. Is it possible to visualize *The Tale of Two Cities* eventually being produced? If so, would it take a year? a million years? a billion years? Of course, we don't know. Most likely, we would conclude that it would never happen. Our experience tells us that random changes introduce disorder and degrade information.

Now imagine the same process happening to a reptile evolving into a bird. Instead of a fast computer, we are dependent on mutations. Mutations are very rare. Many generations might not produce one mutation. When the mutations occur, they are almost always, if not always, detrimental. Natural selection weeds them out as our computer program rejected changes that were not an advantage. How long would it take to evolve a reptile into a bird? If the computer program took many millions of years, how long would it take for just this one major step in the evolution of the vertebrate sequence?

Mathematicians have studied such problems and have concluded that complex systems cannot evolve by pure trial and error means. In 1967, the mathematicians and evolutionary biologists got together to apply such studies to evolution at the Wistar Institute in Philadelphia. Much to the chagrin of the biologists, the mathematicians pointed out that "there is a considerable gap in the neo-Darwinian theory of evolution, and we believe this gap to be of such a nature that it cannot be bridged within the current conception of biology." [quoting Professor Schützenberger in "Mathematical Challenges to the Neo-Darwinian Interpretation of Evolution" as reported in Johnson (1993)]

Scientists have made many calculations of the probability of proteins forming, the probability of DNA forming, and the probability of a cell forming. Some estimates were prepared by evolutionists, some by creationists. These numbers are reported in creationist literature in abundance. Many are inconsistent, being based on different assumptions. Since we can't really understand the numbers (they indicate probabilities unimaginably small), we won't make a big deal about them, but here are a few examples.

> The probability of the chance formation of a hypothetical functional cell, given all the ingredients (a condition far too good to be true) is less than 1 in 10^{57800}. (For comparison, there are only about 10^{80} electrons in the entire universe)
>
> To create a cell by chance, at least one hundred precisely specified proteins must simultaneously appear in one place. The random synthesis of each protein has a probability of approximately 10^{-20}. One hundred simultaneous events yields a total probability of 10^{-200}.
>
> Hoyle and Wickramasinghe estimate the probability of life originating from existing proteins to be 10^{-20}.[26]
>
> Taking into account the requirement to form only peptide bonds with only left-handed amino acids in the correct sequence, the probability of a functional protein forming by chance is 10^{-65}. That is equivalent to correctly selecting one atom out of our whole galaxy.
>
> To evolve fermentative bacteria to fully photosynthetic bacteria involves five new proteins with a random occurrence probability ranging from 10^{-40} to 10^{-104}.

Let's return now to our analogy. In the molecular world where life is built, the analogy to stringing letters and words together is almost perfect. Proteins are strings of

---

[26] Hoyle is a Nobel prize winner who believes this value is so improbable that panspermia represents a more probably cause for the origin of life on Earth.

amino acids; DNA is a string of nucleotides. Of the gazillions of combinations of amino acids and nucleotides, only a few have biological meaning. To find them by chance is less likely than plagiarizing Dickens with improbable random substitutions. But our analogy is not perfect for we were far too generous.

In our example, the changes allowed in the student's text were directed. Only changes that move the text closer to the target were permitted. As any Darwinist will tell you, evolution is not directed, has no endpoint in mind. The watchmaker is blind. The only direction allowed is that which enhances survivability.

We could improve our analogy by removing the artificial intelligence that directed the evolution toward Dickens, thus allowing a greater variety of random changes. Natural selection could be modeled by programming the computer to accept changes that would "survive" the test of the English language. Such a requirement would eliminate words like "qwerty" but allow words like "mitochondria," neither of which would move the discourse toward Dickens. Such a change would considerably slow down the convergence on Dickens' text, but it would be more in line with Darwinism. Under this improved simulation, our evolutionary pathway towards *The Tale of Two Cities* might wander through *Pilgrim's Progress* or even *The Blind Watchmaker* before arriving at its destination.

But the analogy produces results that may still be too probable. Given the complexity of life, the test would be more fair if we changed individual letters instead of whole words. Clearly, no amount of time filled with random processes will compensate for highly improbable events. Random change always degrades information; it never improves it. Evolutionists only have a billion or so years to work with, but even infinite time will not solve the probability problem. Billions of years cannot overcome the argument for intelligent design.

The Intelligent Design Theory Revolution

The Evolution of Creationism

Although there have been "creationists" ever since the beginning of human history, the modern creationist movement began in the early part of the twentieth century with George McCready Price (1870 - 1963). Prior to Price, there were various individuals with anti-evolutionist agendas, but none had a systematic view on the origin of the Earth and none engaged the tools of science in their crusades against evolution. Price, who believed in a literal six-day creation, proposed a flood-geology catastrophism that was new thinking for the time. In 1929, one of Price's former students, the biologist Harold Clark, published a book that formally adopted the term "creationism" to represent Price's flood-geology. The book, *Back to Creationism*, promoted the idea of a science of creationism and criticized pure hostility to evolution without an alternative.

Price's catastrophism remained a minority opinion among creationists until Henry Morris and John Whitcomb published *The Genesis Flood* in 1961. This book remains the quintessential classic of creationist literature. Whitcomb, Morris and others with advanced degrees in science then formed the Creation Research Society in 1963. The CRS was committed to young-earth creationism. The movement has grown since then with more or less unified views on Earth's geologic past. Numerous other organizations and societies have arisen to meet the needs of the growing creationist viewpoint.

Until recently, the creationist movement has concentrated on gaining influence in the public schools. However, in 1987, the U.S. Supreme Court ruled that laws promoting creation science, such as had passed in Arkansas and Louisiana, violated the First Amendment to the Constitution. Unfortunately, the largest audience of the creationists were other believers. Creation scientists almost exclusively published in their own journals, not in the open scientific literature. One notable exception is Dr. Robert Gentry, formerly of Oak Ridge National Laboratory, who published in the most prestigious scientific journals in the world.

The Birth of Intelligent Design Theory (IDT)

The Search for Extra-Terrestrial Intelligence (SETI) originated in 1960 at the National Radio Astronomy Observatory. The program has grown considerably and is now financed largely by NASA. Its crown jewel is the radio telescope at Arecibo, Puerto Rico. SETI has more than its fair share of (shall we say) "visionaries," but it is considered a respectable scientific program. The principal scientific enterprise of SETI is to distinguish intelligent patterns from random noise. SETI, as well as cryptography and forensic science, have developed special techniques for detecting intelligence, based on information theory. No one challenges these endeavors as being unscientific. Therefore, why should anyone challenge biologists who search for indications of intelligent design in living things?

IDT science is just like SETI - a search for intelligent intervention in life. Although there were some precursors in the 1970's, the IDT movement was launched in 1984 with the publication of *The Mystery of Life's Origin* by Charles Thaxton, Walter Bradley, and Roger Olson. Soon after was Michael Denton's *Evolution: A Theory in Crisis*, which has been frequently quoted in this book.

Without appealing to theological arguments, these books, authored by insiders to evolutionary science, provided a most powerful critique of evolutionary theory.

However, it is Berkeley law school professor Phillip E. Johnson who is the acknowledged leader of the IDT movement. Using the tools of logic and rhetoric as only a lawyer can, Johnson not only critiques evolutionary theory, but he also exposes Darwinism as the scientific manifestation of the philosophical worldview of naturalism. "Darwinists play a shell game by getting you to assent to the trivial definition of evolution, and then, suggesting that it compels you to accept a comprehensive philosophy of naturalism," says Johnson in his powerful and influential 1991 book, *Darwin on Trial.*

The next big blockbuster in IDT was Michael Behe's, *Darwin's Black Box.* Behe completed the development of IDT by illustrating how IDT can become a legitimate scientific research program conducted at our most prestigious institutions. Others have produced important works as well, including William Dembski, Stephen Meyer, J. P. Moreland, Paul Nelson, Jonathan Wells and others. What is important is that these individuals are not only publishing to others interested in IDT but in scholarly, peer-reviewed journals as well.

In a nutshell, IDT declares that intelligence is required to explain the irreducibly complex and information-rich systems of life. Furthermore, IDT states that intelligent design is empirically detectable using techniques already in use in other sciences. Therefore, IDT does not appeal to the Bible or to apologetics. IDT does not speculate on the nature of the intelligence. It critiques Darwinism on scientific and philosophical grounds while offering a scientific research program in its place.

So now we have biblically-literal creationism existing simultaneously with deistic or even atheistic IDT[27], both working towards many similar purposes. While creationism has not been very effective in reaching scientists and others in academia, it has been very fruitful as a tool for Christian evangelism. IDT has proved to be remarkably successful in beginning to unseat naturalism as a powerful influence in western culture, but does a nearly godless scientific enterprise have any real impact on eternal issues? Does IDT limit its epistemology by rejecting revealed truth? This subject will be explored more fully in Chapter 8.

Conclusion

Professor Douglas Futuyma of the State University of New York wrote in his textbook, *Evolutionary Biology* (Futuyma 1998):

> Darwin's immeasurably important contribution to science was to show how mechanistic causes could also explain all biological phenomena, despite their apparent evidence of design and purpose. ... but it was Darwin's theory of evolution, followed by Marx's materialistic (even if inadequate or wrong) theory of history and society and Freud's attribution of human behavior to influences over which we have little control, that provided a crucial plank to the platform of mechanism and materialism - in short,

[27] Many IDT proponents are theists working within IDT's theologically minimalist framework.

> of much of science - that has since been the stage of most Western thought.

After considering IDT, the question before us is: will Western thought ever turn back from the purely mechanistic philosophy of naturalism? Probably not, according to MIT physicist and philosopher Thomas Kuhn, unless something better takes its place. Kuhn invented the concept of paradigm shift in his book, *The Structure of Scientific Revolutions* (Kuhn 1996), which describes how science undergoes major theoretical changes. Science is structured, says Kuhn, by conceptual views of nature that define what is acceptable to pursue in investigation and how experimental data are to be interpreted. Every observation and hypothesis must be force fit into the so-called paradigm that is the current tradition of scientific practice, the common body of belief.

When experimental data are uncovered that do not fit the prevailing paradigm, scientists do not suddenly abandon the paradigm. There is great inertia to shifts in paradigms. "Once it has achieved the status of paradigm, a scientific theory is declared invalid only if an alternate candidate is available to take its place." Has something better arisen that can make naturalism invalid? Is it even possible for the paradigm of naturalism, as expressed through evolution, to be unseated? Never has a scientific theory been so strongly entrenched and intertwined with Western thought and even human nature. Only time will tell. Such a reversal in paradigm might not happen until that time when "*at the name of Jesus every knee should bow ... and every tongue should confess that Jesus Christ is Lord.*" Philippians 2:10.

Five

The Attempt to Make Evolution a Fact

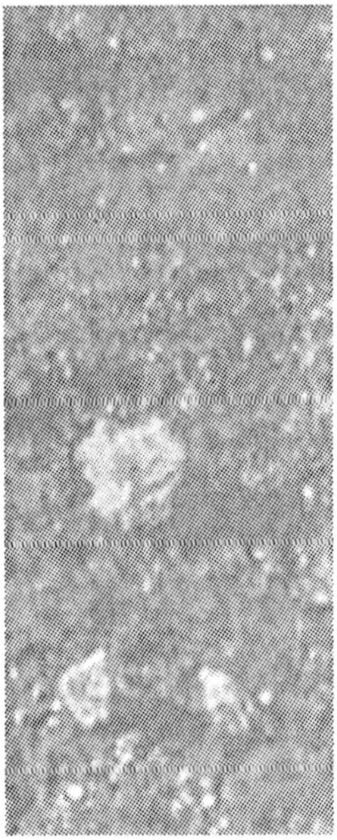

Scientists are known to be objective evaluators of facts - professionals that do not get emotional about their theories, pure seekers of truth. That is the public image, but scientists like Alfred Wegener can tell a different story about the neutrality of scientists in their pursuit of knowledge.

Alfred Lothar Wegener (1880 - 1930) was a planetary astronomer by education but worked primarily in meteorology. He was a real adventurer who enjoyed exploring Greenland and breaking records as a balloonist. During the Great War, he was twice shot and becamc unfit for the battlefield.

Nevertheless, he continued to serve in the German meteorological service where he had time to develop his theory on how the Earth works.

Like Darwin, Wegener's contribution to science was his ability to synthesize a grand theory from dissimilar facts drawn from a variety of fields such as, in Wegener's case, geophysics, geology, paleontology, zoogeography, phytogeography, paleoclimatology, and geodesy. Although others before him had wondered about the apparent geometric fit of the coastlines of Africa and South America, Wegener painstakingly collected all the facts to support a theory that the continents had once all been joined into one supercontinent called Pangaea. In the distant past, Pangaea broke up and the parts drifted apart in a process that came to be known as continental drift.

Some of the evidence that Wegener was able to bring to bear is striking. First of all, the coasts of the continents really do show a remarkable fit (even better at the 1,000-meter depth on the continental shelf as shown much later in 1964). The stratigraphy of eastern North America and northwestern Europe can be matched in great detail, as well as in some of the southern continents. The study of both living and fossilized plants and animals supported the theory.

The opposition to Wegener's theory was intense. One would think that scientists would love a novel theory with so many facts suggesting its veracity. Such a theory would be an exciting puzzle to work on, an opportunity to discuss its pros and cons in conferences and parties, a generator of research programs to pursue. However, that was not the case. The theory went against what the scientists had all been taught. And for shame, Wegener was not one of their own; he was a meteorologist meddling in affairs he should know nothing about.

The great American geologist, Thomas C. Chamberlin, wrote (presumably quoting someone else), "If we are to believe Wegener's hypothesis, we must forget everything which has been learned in the last 70 years and start all over again" (Britannica 2000). If a new theory is based on scientific observation, regardless of whether it were ultimately proved to be true, would it not be a more exciting theory if it turned upside down the existing paradigm? British geologist Philip Lake exclaimed, "he is not seeking truth; he is advocating a cause, and is blind to every fact and argument that tells against it." American paleontologist E. W. Berry characterized Wegener's research as, "a selective search through the literature for corroborative evidence, ignoring most of the facts that are opposed to the idea, and ending in a state of auto-intoxication in which the subjective idea comes to be considered as an objective fact." Geophysicist Harold Jeffreys wrote that the idea of continental drift was, "a very dangerous one, and liable to lead to serious error."

Was Wegener's theory out in left field? Was his evidence weak? Did the nearly unanimous and even excessive opposition to Wegener's theory result from some flaw in his reasoning? Without judging the efficacy of his theory, we would have to say no! Today, Wegener's theory is widely accepted, largely as he proposed it and enhanced by the principle of plate tectonics (even though many questions remain unanswered). The evidence has been bolstered by paleomagnetic studies of the midoceanic ridges and by comparing isotopic ratios of rocks on distant shores.

The reasons for 40 years of indignation against continental drift and Wegener himself are instructive for our current consideration of attempts to make evolution a fact. First, Wegener was not an insider. Human nature does not permit those who have not paid their professional dues to comment on issues for club members only. Secondly, continental drift doesn't sound like uniformitarianism. Thanks to Lyell and Darwin, scientists had just escaped from catastrophism. Their new view of the world did not take kindly to any theory that might retreat from the new geochronology. Finally, Thomas Kuhn has explained in great detail how scientists resist a paradigm shift. The paradigm serves the scientist well in focusing research, but that same focus leads "to an immense restriction of the scientist's vision and to a considerable resistance to paradigm change. The scientist has become increasingly rigid." (Kuhn 1996)

Wegener's theory had little to offend philosophical sensitivities. Its connection to catastrophism was tenuous. Yet the scientists' classic resistance to change brought almost violent protest against a scientific proposal that should have been coolly examined and evaluated. The history of resistance to continental drift provides lessons for creationists today. In the case of creationism, all the factors that led to Wegener's difficulties are present, plus the challenge to beliefs on some of the most fundamental issues mankind faces. Under these circumstances, should we expect the intelligentsia of today to act more nobly than those in Wegener's day? Certainly not, especially in light of the fact that, until Gentry presented geochronological evidence for instantaneous, fiat creation and the IDT revolution offered evidence of design in nature, creationists (or IDT proponents) had offered very little in the way of alternatives to evolution.

This chapter describes some of the negative reactions to creationism offered by evolutionary scientists and their attempts to proclaim evolution as a fact, even though it has not undergone the process historically reserved for theories to become "facts."

Laws, Theories, and Facts

As scientists observe nature, they are often able to recognize regularities existing in objects or processes that seem to always hold true. Although the scientist may not be able to understand the phenomenon, he knows the outcome of an observation to be the same every time it is observed. Such generalizations can eventually be categorized as scientific laws. An example is Newton's laws of motion. He was able to express the pattern of motion in three laws that he could mathematically define with equations. However, Newton had no understanding of why the laws appeared to hold true. Some other scientific laws are the ideal gas law, the law of electrostatics, the law of conservation of energy, and the law of biogenesis.

A scientific theory is a proposed explanation for observed phenomena that arises in the context of a law and results from the creative but rational mind of the scientist. As such, the theory is subject to human weaknesses such as bias from the scientist's philosophical worldview and the limitations of his particular scientific experience. Theories often arise as hypotheses (preliminary, untested ideas) and evolve into theories as confirming evidence mounts.

Scientists test theories by either continuing to observe nature in light of the theory or by intervening in nature with stimulations to observe or measure the effects. Such experimentation may prove the theory to be false. But as empiricist philosopher David Hume proposed, only an infinite number of confirming tests can prove a theory to be true. Therefore, theories have difficulty moving from proposition to fact. Repeated substantiating experiments may establish a continually improving confidence that the theory is true, but the theory is falsifiable by only one observation. Philosopher of science,

Karl Popper (1902 - 1994) argued that scientists should establish falsifiability criteria and seek experiments to falsify their theory's - not merely seek corroborating evidence. The failure to falsify is greater evidence of the theory's reliability than the confirming experiment. Popper states that so-called theories that are not falsifiable (e.g., Marxism, Freudism, astrology) are pseudosciences (Popper 1989).

A fact is not a formal construct in science; the word is used in science just as it is used in everyday life. Perhaps we could view facts as very small laws. A theory is an explanation for a body of facts. It is a fact that objects falling near the Earth's surface increase their velocity at 9.8 meters per second every second. This could be stated as a law, but it is a single fact that subjectively doesn't seem to warrant law status (although Newton's second law effectively captures the concept but not the magnitude). Facts are assumed to be true, but everyone's human experience reveals that facts are not always what they seem.

In spite of their self-evident nature, laws, like theories, can be shown to be false. The law of conservation of energy had to be reformulated when it was learned that mass and energy are just two forms of the same entity. Newton's laws of motion were found to be special cases of Einstein's special theory of relativity. Also, theories can be graduated to law or fact status. For example, the Copernican theory of the Earth rotating around the Sun has been observed so unambiguously that its truth is not questioned at all; it is a fact. However, a theory graduating to fact or law requires considerable empirical support and must be unanimously attested to by all rational observers.

In light of this understanding of the relationship of theory to fact, what are scientists to think when evolutionary philosopher Michael Ruse exclaims, "Evolution is a fact, fact, FACT!" (Ruse 1982). As we have seen in the preceding chapters, evolution does not satisfy the criteria for a fact; it hardly qualifies as a theory. The falsifiability of evolution is even in doubt. It has never been observed to occur (except for microevolution, which doesn't count). There have been no credible mechanisms suggested for transition from one species to another. It defies the observed typological nature of life. It is inconsistent with the irreducibly complex architecture of living systems. There is far from universal acceptance of evolution by qualified, rational scientific investigators. How then can it become a fact?

In Professor Futuyma's textbook on evolution (Futuyma 1998), he tries to distinguish the fact of evolution from the theory of evolution, "evolution is a scientific fact. But it is explained by evolutionary theory." This artificial distinction is unconvincing. His book is a thorough presentation of the theory of evolution, but it does not establish the fact of evolution. Interestingly, Futuyma also says this about evolution and facts:

> In contrast to fields such as physiology and biochemistry, in which many specific, detailed facts are an essential complement to the field's broad principles, evolutionary biology encompasses so much, and evolutionary biologists engage in such diverse studies, that the number of specific facts that all persons trained in evolution should know is rather limited.

Is Futuyma's rationale for the lack of facts convincing? Is evolutionary biology really more diverse and encompass so much than biochemistry or physics? Or is there a lack of facts and an abundance of storytelling. In a scientific field in which the main process has never been observed to occur, one would expect a paucity of facts.

Other scientists have tried to convince us that evolution is a fact, without letting the facts of evolutionary science speak for themselves. No other scientific discipline has been the subject of such a public relations campaign. Examine these statements.

> The first point to make about Darwin's theory is that it is no longer a theory but a fact. ... Darwinism has come of age so to speak. We are no longer having to bother about establishing the *fact* of evolution .... [Julian Huxley in 1959 as reported in Denton (1985)]

> The theory is about as much in doubt as the earth goes round the sun. [Dawkins in The Selfish Gene as reported in Denton (1985)]

> Evolutionary biologists today do not concern themselves with trying to demonstrate the reality of evolution. This is simply no longer an issue, and hasn't been for more than a century. (Futuyma 1998)

> However, the historical reality of evolution - the descent, with modification, of all organisms from common ancestors - has not been in question, among scientists, for well over a century. It is as much a scientific fact as the atomic constitution of matter or the revolution of the earth around the sun. (Futuyma 1998)

These statements are so demonstrably wrong that one wonders why scientists would be willing to go into print with them. They believe in them so we should not cite them for dishonesty. However, the enthusiasm for the campaign for evolution sometimes gets out of hand, as evidenced by the following radical statements.

> It is absolutely safe to say that, if you meet somebody who claims not to believe in evolution, that person is ignorant, stupid or insane (or wicked, but I'd rather not consider that).[28] [Dawkins as reported in Johnson (1993)]

> If you insist on teaching your children falsehoods - that the earth is flat, that "Man" is not a product of evolution by natural selection - then you must expect, at the very least, that those of us who have freedom of speech will feel free to describe your teachings as the spread of falsehoods, and will attempt to demonstrate this to your children at our earliest opportunity. Our future well-being - the well-being of all of us on the planet - depends on the education of our descendants. [Daniel Dennett in *Darwin's Dangerous Idea* as reported in Johnson (1998)]

Intimidation such as this is very effective in preventing people from admitting their skepticism of evolution. The intimidation is also applied to scientists that dare to breathe a doubtful word about evolution. A united front must be maintained, even at the expense of accuracy and proper scientific decorum.

---

[28] It is amusing that Dawkins then went on to say that creationists were intolerant.

Intimidation of Scientists

The most famous case of intimidation within science is the trial of Galileo Galilei in 1633. At the age of 69, this highly distinguished scientist was forced, under the threat of torture and of being burned at the stake, to recant on his knees a lifetime of scientific work. Because he was willing to "abjure, curse, and detest" the idea that the Sun is the center of the universe and that the Earth moves around the Sun, he was allowed to continue working under house arrest until his death in 1642.

Galileo had proposed a scientific theory that was not in accordance with the accepted views of the intellectual establishment. The most powerful institution of his day, the church, brought him to the Inquisition on the issue. Today, there is no institution as powerful as that which Galileo faced (except perhaps in very localized situations) and there is no Inquisition to force ideas on people with the threat of death. Nevertheless, the public is lobbied and intimidated as are scientists who do not toe the line on accepted theories. Creationists could be expected to lobby the public (and some even force the public with attempted laws), because they are fulfilling one of the requirements of their philosophical worldview, religious evangelism. However, when evolutionary scientists lobby and intimidate, such actions are contrary to their philosophical worldview. It is not a tenant of the scientific enterprise to propagate truth by manipulating public opinion.

In Chapter 2, we saw how hatefully evolutionists responded to Richard Goldschmidt's "hopeful monster" theory of evolution. Goldschmidt was an evolutionist just proposing a non-Darwinian method. In turn, Stephen Jay Gould, a very prominent and vocal evolutionist of our time, has been vilified by Darwinian evolutionists for proposing the somewhat non-Darwinian punctuated equilibrium. But the real punishment of scientists, the rejection of tenure, the loss of jobs, the termination of funding, is reserved for scientists who are not evolutionists at all.

As described in his 1986 book, *Creation's Tiny Mystery*, physicist Robert Gentry experienced the full force of intimidation and retribution when he testified on the side of creation at the Arkansas trial regarding the state law on balanced treatment of evolution and creation. Gentry was a highly respected scientist at the Oak Ridge National Laboratory, performing work on issues that gained the interest of the world's leading scientists, including Nobel laureates. He published his continually unfolding results in leading scientific journals, including the two most prestigious journals in the world, *Science* and *Nature*.[29]

After the trial, the scientific press misrepresented Gentry's testimony and did not allow him the opportunity to correct the record. Oak Ridge National Laboratory, after having evaluated his work to be of value for twelve years, suddenly terminated the invitation for him to work at the Laboratory. Did Gentry's work become scientifically suspect? Of course not; it was his beliefs that were unacceptable, the conclusions that he drew from this published scientific research that were not in accordance with the accepted philosophical worldview.

The case of Forrest Mims is another example that received nationwide attention. Mr. Mims was a well known science writer with 70 books and several hundred articles when, in 1988, he began negotiating with

[29] Gentry's work will be discussed in Chapter 6.

Scientific American to be the writer for the famous column, "Amateur Scientist." During the interviews with the magazine, it was uncovered that Mims had written for Christian magazines and did not accept Darwinian evolution. Even though Mims assured the editor that the column would not be used to promote Christian beliefs, he did not get the job.

Roger DeHart is a biology teacher at Burlington-Edison High School near Seattle. For 10 years, DeHart taught both the evidence for and against evolution - a position that embodies the open inquiry and self-criticism of science. He was a popular teacher among both students and administration. However, following a student complaint, the ACLU accused DeHart of teaching creation (although students testified that he did not talk about God or religion in class). As a result, Mr. DeHart was ordered to cease teaching students about intelligent design, but he was allowed to continue discussing problems with Darwinism.

More recently, in May 2000, DeHart wanted to present some of the more recent evidence about evolution that was not supportive of Darwinian evolution. He had intended to use the scientific journals themselves, not creationist literature, as his source to correct misinformation in his outdated textbook. However, he was told that he could not present the more recent information. History shows that Darwinists would rather teach false or misleading information than allow the façade to develop any cracks.

For every one of these prominent cases, there are hundreds of others where dissent is stifled and perhaps thousands of others where scientists are just being careful to keep their doubts about Darwinism to themselves. "Evolution is a fact, fact, FACT!" remains the slogan of the Darwinists.

Institutional Pressures

There are a number of institutions that are players in the creation-evolution controversy. Several creationist organizations promote creation science while others seek to propagate creationism in the public schools system. Since these bodies are as much religious institutions as they are scientific ones, they are of no interest to our current purpose of examining intimidation of scientists. They exist to evangelize, making no pretensions to objectivity. On the evolutionist side, there are the National Center for Science Education (NCSE), the American Association for the Advancement of Science (AAAS), the National Academy of Sciences (NAS), and the National Association of Biology Teachers (NABT). We will only briefly mention the first two since they are less interesting. However, the last two will be discussed in more detail.

The National Center for Science Education is an organization dedicated to making creationists look bad. It works hard to counteract the influence creationists have occasionally gained in public schools. NCSE takes extreme positions at times. In spite of the many strong statements emanating from NCSE, it is not of interest to our discussion, because it is an evangelistic organization - not one influential in the actual pursuit of science.

As the world's largest association of scientists, the American Association for the Advancement of Science is a prestigious scientific institution. It is the publisher of the peer-reviewed journal, Science, which is one of the most widely read and respected scientific journals in the world. AAAS has an interest in science and public policy and sponsors papers on the subject, including issues in religion and creation-evolution. Although very much on the side of evolution, AAAS has avoided setting guidelines for teaching evolution and has restrained itself from vitriolic attacks on creationism. AAAS maintains, as do most evolutionists, that evolution is religiously neutral, operating in a distinctly different sphere than religion.

National Academy of Sciences

The most prestigious scientific organization in the U.S. is the National Academy of Sciences. Chartered by Congress in 1863, the NAS is a non-profit, private organization that was founded to advise the federal government on scientific and technical matters. As such, the NAS is duty-bound to reflect the highest principles of science, including open and objective pursuit of scientific truth regardless of nonscientific pressures to support a prevailing view. Membership in the Academy is by election, with only the best and the brightest being eligible. One of the interest areas of the NAS is science education.

In 1976, NAS adopted a resolution on freedom of inquiry and expression that is partially quoted here (Gentry 1986):

> ...That the search for knowledge and understanding of the physical universe and of the living things that inhabit it should be conducted under conditions of intellectual freedom, without religious, political or ideological restriction.
>
> ...That all discoveries and ideas should be disseminated and may be challenged without such restriction.
>
> ...That freedom of inquiry and dissemination of ideas require that those so engaged be free to search

> ...publish...without fear of retribution in consequence of unpopularity of their conclusions. Those who challenge existing theory must be protected from retaliatory reactions.

Does the NAS operate by the principles in its resolution?

In 1984, NAS produced a pamphlet called *Science and Creationism: A View from the National Academy of Sciences.* The pamphlet states, "the Academy states unequivocally that the tenets of 'creation science' are not supported by scientific evidence." While the volume of material that might be classified as 'creation science' is not large compared to that for evolutionary science, it is just simply not true that science does not provide evidence for intelligent design or divine creation.[30] Suppression of evidence that does not fit the popular philosophical position is not consistent with the NAS resolution.

In the Preface to the 'View,' Dr. Frank Press, President of the Academy, states, "The theory of evolution has successfully withstood the tests of science many, many times. Thousands of geologists, paleontologists, biologists, chemists, and physicists have gathered evidence in support of evolution as a fundamental process of nature." Are we to conclude from this that the veracity of a particular theory is dependent upon the number of scientists who support it? Is scientific truth determined by vote? The more appropriate approach is to allow inquiry, consistent with the NAS resolution, and then

[30] The very term, Creation Science, although invented by the creationists themselves, gives the impression that there is some body of work for which its sole *raison d'être* is to support creationism. It would be better to look upon scientific inquiry as a quest for truth without any preconceived outcome. Scientific study should uncover phenomena; the interpretation of the phenomena may be consistent with a creationist or evolutionist bias.

provide for critical peer review of the methods and results. Issuing proclamations of support for a theory based on the number of scientists that support it is not scientific and is not in accordance with the NAS resolution.

In 1998, the NAS prepared another book, *Teaching Evolution and the Nature of Science.* The book's purpose is to help teachers and publishers to resist anti-evolutionist influence. Again, the authoritarian means of establishing truth are employed:

> Those who oppose the teaching of evolution in the public schools sometimes ask that teachers present "the evidence against evolution." However, there is no debate within the scientific community over whether evolution occurred, and there is no evidence that evolution has not occurred.

Besides asking creationists to do the nearly impossible, to prove a negative, the statement is just plain false. There is indeed debate within the scientific community about whether evolution has occurred. There would be a lot more debate if the NAS would promote freedom of inquiry and dissemination of ideas without fear of retribution.

The book further states, "It is no longer possible to sustain scientifically the view that the living things we see today did not evolve from earlier forms or that the human species was not produced by the same evolutionary mechanisms that apply to the rest of the world." The authors try to support this misstatement by several lines of evidence, but the evidence offered is the same tired extrapolation of microevolution to macroevolution.

Remarkably, the NAS, an organization without interest in religion, attempts to define for religious people what religion means, "At the root of the apparent conflict between some religions and evolution is a misunderstanding of the critical difference between religious and scientific ways of knowing. Religion and science answer different questions about the world." This separation of religion and a systematic approach to truth via the scientific method is reminiscent of Bacon's ideas of mutually exclusive revealed truth and natural truth.

However, as much as evolutionists would like to separate religion and science into two separate endeavors, religion is very much interested in science and the natural world and also uses systematic means to derive truth from both revelation and nature. Evolutionists would have the public believe that evolution is religiously neutral, but this is simply not true. In spite of numerous statements to the contrary, evolution denies the existence of divine influences. Cornell University professor William Provine calls for a more intellectually honest stance among evolutionists. He describes the religion of evolution as totally atheistic and mechanistic (Johnson 1993):

> Modern science directly implies that the world is organized strictly in accordance with mechanistic principles. There are no purposive principles whatsoever in nature. There are no gods and no designing forces that are rationally detectable....
>
> Second, modern science directly implies that there are no inherent moral or ethical laws, no absolute guiding principles for human society.

> Third, human beings are marvelously complex machines. The individual human becomes an ethical person by means of two primary mechanisms: heredity and environmental influences. That is all there is.
>
> Fourth, we must conclude that when we die, we die and that is the end of us ....
>
> Finally, free will as it is traditionally conceived - the freedom to make uncoerced and unpredictable choices among alternative possible courses of action - simply does not exist. ...There is no way that the evolutionary process as currently conceived can produce a being that is truly free to make choices.

Thus, we see that the most prestigious scientific body in the nation feels the need to intervene in one area of science and make proclamations rather than allowing the self-correcting scientific process work its way toward truth. Unfortunately, the intervention is fraught with inaccuracies that appear as if intellectual freedom is being repressed.

National Association of Biology Teachers

The NABT is an organization dedicated to enhancing excellence in the teaching of biology. NABT got into trouble when in 1995, it issued a "Statement on Teaching Evolution" to assist high-school teachers in dealing with evolution. The statement contains numerous dogmatic statements that would be interesting to disassemble. However, for the sake of brevity, we will discuss only one.

Although it has grown fashionable for evolutionists to eschew the atheistic position that is central to evolution, NABT seemed to have forgotten the new marching orders. The 1995 Statement read, in part (Johnson 1998):

> The diversity of life on earth is the outcome of evolution: an unsupervised, impersonal, unpredictable and natural process of temporal descent with genetic modification that is affected by natural selection, chance, historical contingencies and changing environment.

The terms "unsupervised" and "impersonal" revealed the inherent atheistic bias of the NABT. If science is really mutually exclusive of religion, how can it draw conclusions about whether God exists? The NABT statement endorsed the religion of atheism, something that even science is supposed to not be able to do. Worse, as the Supreme Court has so carefully explained to the creationists, public schools cannot promote a religion - even the religion of atheism.

Recognizing their error, NABT, with strong encouragement from the National Center for Science Education, revised the statement to remove the religious bias. Of course, this created media attention and ammunition for the creationists. In attempting to impose atheistic views on the nation's high schools, to fill the evidence vacuum with proclamations about the validity of evolution, the NABT reaped a whirlwind of trouble. Intelligent design theory received nationwide attention in the op ed columns. Intellectual opinion, as influenced by the news media, moved one notch closer to sympathy with intelligent design and away from authoritarian evolution.

Indoctrination Through Textbooks

Perhaps one of the most distressing activities undertaken to ensure that evolution is regarded as fact is the perversion of high-school text books. One of the most colorful examples is the following:

"You are an animal, and share a common heritage with earthworms and dinosaurs, butterflies and sea stars." *Biology, Visualizing Life*, Johnson; Holt, Rinehart, Winston, 1994, page 453.

This declaration is made without the slightest hint that evolution is a theory that has not been demonstrated to actually have occurred. Other examples include the drawings of embryos that support the recapitulation concept (ontogeny recapitulates phylogeny - see the next subsection). It has been known for decades that the original figure on which subsequent ones were based was actually fudged to be more in accordance with expectations of the recapitulation concept. Of course the poor old moth is presented again to be an example of "evolution in action," but we should hardly expect high-school students to understand that industrial melanism does not represent evolution when the Ph.D. biologists keep forgetting that point.

In 1995 the Alabama State Textbook Commission performed a periodic evaluation of biology textbooks and found the following types of flaws among those they examined (Anderson 1995):

- naturalistic theories of origins taught as fact

- untestable philosophical assumptions presented under the color of scientific knowledge

- the meaning of "theory" redefined

A Message from the Alabama State Board of Education

This textbook discusses evolution, a controversial theory some scientists present as a scientific explanation for the origin of living things, such as plants, animals and humans.

No one was present when life first appeared on earth. Therefore, any statement about life's origins should be considered as theory, not fact.

The word, "evolution" may refer to many types of change. Evolution describes changes that occur within a species. (White moths, for example, may "evolve" into gray moths.) This process is microevolution, which can be observed and described as fact. Evolution may also refer to the change of one living thing to another, such as reptiles into birds. This process, called macroevolution, has never been observed and should be considered a theory. Evolution also refers to the unproven belief that random, undirected forces produced a world of living things. There are many unanswered questions about the origin of life which are not mentioned in your textbooks, including:

Why did the major groups of animals suddenly appear in the fossil record (known as the Cambrian Explosion)?

Why have no new major groups of living things appeared in the fossil record in a long time?

Why do major groups of plants and animals have no transitional forms in the fossil record?

How did you and all living things come to possess such a complete and complex set of "instructions" for building a living body?

Study hard and keep an open mind. Someday you may contribute to the theories of how living things appeared on earth.

- problems unexplained by evolutionary theory not mentioned
- "evolution" not defined with precision - a classic shell game with words
- biased treatment of environmental problems and social issues
- intellectual dishonesty

The State needed a textbook but could not find one that did not misrepresent the scientific method and the theory of evolution in particular. Therefore, Alabama felt the need to place a paste-in disclaimer in the cover of one of the books and use it anyway. The disclaimer is provided in the text box found in this subsection. (Other states have also reacted negatively to the evolutionist propaganda campaign.)

A tiny sampling of the shorter and more interesting quotes from the books are the following:

> Evolutionary change is undeniable...For example, there is no question that if you jump into the air, you will end up on the ground below. It makes no difference whether you understand - or even believe in - gravity. What goes up must come down. Just as definitely, life on Earth evolves, or changes over time." *Biology*, Miller and Levine, 1995.

Every word of this statement is true, obvious to anyone. But is it just as definite as gravity that all life evolved from a common ancestor? That is the logical jump the authors wish the unsophisticated readers to make. This is another instance of playing the shell game with the meaning of the word, evolution.

> The fossils of the earliest birds are rare, often poorly preserved, and very similar to those of many small dinosaurs. Because of this, there is much controversy over which fossils are those of birds and when birds first appeared on Earth. Although the fine points of bird evolution are hotly debated, one thing is certain - birds evolved from ancient reptiles. *Biology*, Miller and Levine, 1995.

> Common usage of the word 'theory' suggests a guess or an uncertainty... The opposite meaning is true in science. A theory is as close to a complete explanation as science can offer. *Biology Living Systems*, Raymond, Oram, Gencoe, 1994.

The placement of the paste-in by the Alabama Board of Education evoked a furious outburst from evolutionists and the media. The press made frequent allusions to "Inherit the Wind," the film depicting the Scopes trial that portrayed the creationists as ignorant, religious bigots. However, Alabama's action was not a science versus religion issue but a science versus philosophy issue, with the State of Alabama on the side of science. Integrity in education should insist in a clear distinction between fact, theory, and philosophy. Apparently, current biology textbooks are lacking in this respect.

Famous Hoaxes of Evolution

We must be careful in approaching this subject, because of the danger of imputing dishonesty to currently working evolutionists. The great majority of scientists are honest and dedicated to the pursuit of knowledge, including evolutionists. Nevertheless, it is instructive to consider to what lengths humans are willing to go to propagate a worldview that is dear to them. Also, we find that the hoaxes described below did "not go gentle into that good night" when they were discovered. The occasional textbook or documentary today continues some of these hoaxes. Certainly, the generations of scientists trained to believe in them were affected by them.

Most of the following examples are of actual fakes and frauds. In some cases, the scientists themselves were not the perpetrators, having been deceived themselves. There are many other examples that were honest mistakes in which there was a failure to correct them in a timely manner, allowing the false data to be used in textbooks. Such examples could be construed as fraud. The intent of this section is not to ridicule scientists who made a mistake, but to illustrate how quickly scientists see what they expect to see - especially when recognition is the reward.

Recapitulation

In 1866, Ernst Haeckel (1834 - 1919) proposed his biogenetic law in which "ontogeny recapitulates phylogeny." Ontogeny is Haeckel's word for the embryological development of the individual organism, and phylogeny (also a Haeckel invention) is the evolutionary history of a species. Thus, in simple words, as an embryo develops, its morphology successively passes through the adult forms of all its evolutionary ancestors.

Darwin actually thought of it first, "The community in the embryonic structure reveals community of descent."

Haeckel produced a number of books which contained illustrations of several species at three different stages of development, showing the remarkable similarity at each stage. However, Professor Wilhelm His (1831 - 1904), a comparative embryologist, showed that Haeckel had fabricated some of the drawings and had modified others to fit his "law." In 1958, Sir Gavin de Beer of the British Natural History Museum commented, "Seldom has an assertion like that of Haeckel's 'theory of recapitulation', facile tidy, and plausible, widely accepted with critical examination, done so much harm to science." (Taylor 1991)

A scientific concept that was demonstrated to be false (and even dishonestly propagated) as far back 1874 and denounced numerous times since then should have disappeared from the annals of science. However, Haeckel's deliberately forged drawings have continued to appear in biology text books. *The Illustrated Origin of Species*, Abridged & Introduced by Richard E. Leakey, which was published in 1979, contains Haeckel's drawings. For over a hundred years, biologists knew that Haeckel's drawings were partial fabrications; however, students over that time have been led to believe that recapitulation is supported by embryology.

The fact of the matter is that embryologists have thoroughly repudiated the biogenetic law. The law has been replaced by de Beer's law that simply states that for a given species, embryonic features of higher taxa appear before the features of the lower taxa. That is, all embryos of a given taxonomic level look pretty much the same and grow increasingly diverse. For example, the so-called gill slits in human embryos become parts of the face and never have any part in respiration, whereas the analogous structures in fish embryos do finally become gills.

Recapitulation will not die. It has become part of the lore of biology, having been taught to too many generations of rising biologists. The perceived need to cling to the "proofs" of evolution, even after they are disproved, is evidence of the desperation to maintain the façade.

Piltdown Man

Amateur fossil hunter Charles Dawson found his way into the history books in 1908 when he found a skull fragment and most of a jaw in the gravels of Piltdown near Charles Darwin's home town. Dawson carried his find to a leading paleontologist at the British Natural History Museum, Arthur Woodward. Woodward called in the distinguished Arthur Keith, a physician and professor of anatomy, and Grafton Smith, a renowned brain specialist. These men were later joined by Pierre Teilhard de Chardin, a Jesuit priests and scientist, who found a tooth in the same locale.

Although there were some outside this circle who thought that the jaw and skull did not belong to the same species, and there were some who thought the tooth had file marks on it, this group of scientists, the best in the world for this subject, was convinced they had found the missing link. This discovery made a huge impact in scientific circles as well as to lay people. Opposition to evolution, especially human evolution, was greatly reduced. For decades, the textbooks told the story of man's evolution with the indisputable evidence of Piltdown Man providing the support.

However, in 1953 the hoax was discovered.[31] The skull fragment was a modern man. The jaw was that of an ape. Strategic parts had been removed to confuse investigators into thinking the parts were from the same individual. The tooth was filed down to provide conclusive evidence of ancient man. Who perpetrated the hoax? No one knows, but the two suspects are Father de Chardin and Sir Arthur Conan Doyle. This episode is perhaps the most famous fraud in all of science history.

Soon after the hoax was announced, it became obvious by cursory examination that the bones were a hoax. However, these eminent scientists, men of integrity (except perhaps one), had such visions of discovery and expectations to find half-men, half-apes that they saw what they wanted to see. Do anthropologists of today suffer the same weakness?

Charles Dawson's short-lived reward in all this was a hominid species named after him, *Eoanthropus dawsoni*.

Nebraska Man

The case of Nebraska Man was not initially an intent to deceive. Nevertheless, considerable dishonesty resulted from it at a particularly inopportune time for creationists. In 1922 a single tooth was found in Nebraska. Professor Henry Osborn, head of the American Natural History Museum, determined it to be from a species representing the "missing link." Grafton Smith, the same man of Piltdown fame, convinced the *Illustrated London* News to run an artist's rendering of what the Nebraska Man must have looked like. Although the artist only had a tooth, a double page spread showed a naked ape-man

---

[31] The author recalls as a child reading about Piltdown Man in science books in the 1950s, perhaps 1956. Although the books were likely written before the hoax was discovered, such concepts do not easily disappear from the mythology of evolution.

of imagined characteristics. The picture received world-wide distribution.

Just three years later, the famous Scopes trial took place in Dayton, Tennessee, in which the case for creationism was argued in the courts. The recent discovery of Nebraska Man was clear in the minds of those present and was used to make the creationists look foolish. After the trial, it was discovered that the tooth actually came from an extinct pig[32]. No correction of the error was announced in the public venue. The Nebraska Man was quietly removed from the encyclopedias and text books. Scientists with an agenda lose the ability to be objective and find it easy to misrepresent the truth. If aggressive evolutionists can develop a whole race of people from a pig's tooth, what can they do with a complete skeleton of an extinct primate?

Archaeoraptor

The most recent evolution hoax was uncovered just this year (2000). In February 1999, private museum curator, Stephen Czerkas, found a fossil at a fossil fair in Tucson that consisted of a bird with a dinosaur tail. The fossil, named *Archaeoraptor* (not the same as *Archaeopteryx*), came from China. Although *National Geographic* was warned ahead of time that the fossil may be a hoax, the magazine published pictures of the fossil, and held a press conference that proclaimed the find as important evidence for reptile-to-bird evolution. This news was so important that the find was given national attention in the open press.

---

[32] Eventually, the pig was found to be not extinct but still living in Paraguay.

Chinese scientists looking into the find discovered that farmers in the region where the fossil was found have learned the art of fabricating mosaic fossils. Fossil making is apparently a significant source of revenue for the region. Although both *Science* and *Nature* rejected a paper by Czerkas on the subject, the lure of fame and the enthusiasm to support a cherished theory often leads scientists to see what they want to see. Unfortunately, more than scientists saw the fossil; nine million school children saw it in *National Geographic*'s Explorer's Hall before it was removed.

Scientists should not be condemned for making a mistake. Science proceeds by halting movement because hypotheses are proposed and then carefully tested. Sometimes the hypothesis is wrong; sometimes the experimental data are wrong. However, the greatest wrong is when scientists rush to judgement without the careful investigation, self-critical evaluation, and collaboration and consulting that normally constitutes good science. Holding a press conference at the first opportunity does an injustice to the name of science.

Archaeopteryx is another fossil creature that is clearly bird but has some reptile features. *Archaeopteryx* has long been one of the jewels of evolutionary theory - the one fossil that indicates a transition according to theory. At one point, Nobel prize winning scientist, Fred Hoyle, declared *Archaeopteryx* to be a hoax. However, in recent years, more specimens have been found. Most scientists now do not believe that Archaeopteryx is a hoax. *Archaeopteryx* is discussed in Chapter 7.

Conclusion

Propaganda, distortions of truth, and even the occasional hoax are characteristics of how the science of origins is conducted, both by the creationists and the evolutionists. Perhaps we should forgive the creationists (at least for propaganda), because they are truly evangelists for their cause. Besides, the intelligentsia have locked them out of the debate, leaving them no recourse but to make points the best ways they can.

However, the evolutionary scientists have demanded that they have the moral high ground, that they alone are on the side of science. We would then expect from them objectivity and an openness to new ideas as embodied in the NAS resolution, a pursuit of science with the scientific method instead of the lawyer's tools of debate. Then why do we see the behavior described in this chapter?

First, as has been noted as early as Chapter 1, we are not so much dealing with scientific issues as we are philosophical ones. The religion of evolutionists, and much of the intellectual world, is naturalism. Religious beliefs are often defended to the point of death.

Secondly, creationists are attacking evolution. (One must wonder if creationists using science to attack evolution is any different than the evolutionist conducting science to prove a preconceived notion.[33]) However, what frightens evolutionists the most is that creationists are attacking evolution in the nation's schools. Evolutionists seem to think that all of science and per-

[33] The author has no argument with creationists as scientists. However, science of origins should be conducted in a manner of discovering what the Creator has prepared for us, not seeking to prove a preconceived idea. One will always find "proof" of a cherished idea.

haps the entire epistemology of modern times will be eroded by this. But the truth is that nearly all physicists can conduct their inquiries irrespective of what evolution teaches. Nearly all chemists are unaffected by evolutionary theories. Even most biologists would not miss a beat in their research should evolution be suddenly and unequivocally disproved, in spite of Dobzhansky's famous quote, "Nothing in biology makes sense except in the light of evolution."

Finally, evolutionists are not winning the propaganda wars. A recent Gallup poll indicates that 44 percent of all Americans believe in the biblical view of origins, including creation of humans less than 10,000 years ago. Only 10 percent believe in a purely mechanistic evolution. When the questions are only asked of scientists, 5 percent believe in the biblical view and 40 percent believe in theistic evolution. Amazingly, only 55 percent believe in mechanistic evolution. Thus, the high priests of evolution, the ones whose education and careers are based on evolution, are conducting desperate propaganda wars. They have succeeded among the intelligentsia but have failed among the general public.

However, if scientists are so concerned about the integrity of science and the state of education in America, why do they resort to propaganda and distortion?

Six

Did Lyell Get It Right?

Issues in Geology

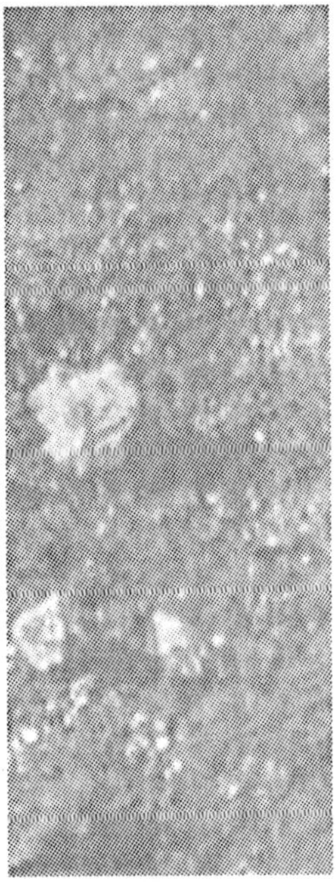

On May 18, 1980, Harry Truman learned a lesson on catastrophism - not President Harry Truman (Jimmy Carter was President at the time) - but 83-year old Harry Truman, proprietor of the lodge on Spirit Lake. Harry had moved to the lake in 1926 to avoid troubles with gangsters; he had been a bootlegger. For 54 years, he had enjoyed the peace of his lake and the mountain that shadowed it. Therefore, Harry knew the mountain well. Even though everyone from the geologists, to the local authorities, to the news media had warned him to move away, Harry believed that his mountain would not cause him any trouble. Without knowing the meaning of the word, Harry believed in uniformitarianism, at least his version of it - all things would continue as they always had. Mr. Truman would not move away.

Scientists believe that just before 9 am on that May morning, Harry had just enough time to turn his head before a 300-miles-per-hour avalanche of searing hot gases, mud, debris, and molten rock buried him and Spirit Lake 150 feet deep. The eruption of Mt. St. Helens displaced approximately 0.6 cubic mile of rock, replacing the conical shape of the volcano with a horseshoe shaped crater. The blast utterly devastated 210 square miles. The U.S. Forest Service estimated that 10 million mature trees were felled, with many more killed.

Harry never benefited from his several millisecond long lesson in catastrophism, but today geologists are reaping volumes of data on how many of Earth's geologic structures can be formed in just a few hours - contrary to previous theories. Catastrophism is making a comeback. This chapter presents some of the evidence for a catastrophic origin of many of the Earth's features and addresses some of the evidence for the assumed long age of the Earth, which has been a prerequisite to biological evolution.

The Message of Mt. St. Helens

The aftermath of the Mt. St. Helens eruption has presented geologists with an opportunity to see how strata that apparently took millions of years to form can be deposited in a matter of hours. Amazingly, there are 30 feet of deposits at Mt. St. Helens that are finely layered as if hundreds or even thousands of separate events deposited them over millions of years. Even creationists were astounded to see so many layers deposited in a roughly nine-hour event. Geologists would have expected to see one large layer or one that is a continuum of change.

These deposits were then rapidly eroded away as the Toutle River, which had been dammed by a mud flow, cut through the dam and washed over the deposits. The deluge cut a deep canyon that looks very much like the Grand Canyon but 1/40th scale. Geologists were already beginning to doubt that the Colorado River carved the Grand Canyon. Now they have a model proven to be effective in creating wide, steep-cliffed canyons.

The new Spirit Lake, which has formed 200 feet higher than the original one, is filled with about a million logs that were washed into the lake by an 860-foot high wave that scoured the mountainside. Although most of the logs are floating prone, many are floating vertically. (The root end of trees is more dense, causing the root end, in some cases, to sink.) As these logs become waterlogged, they sink to the bottom where they begin to be covered by silt. Since the logs are sinking over years of time, the root end of the logs are buried in the sediments at different layers. This conditions offers a proven means of creating formations like the so-called multiple forests of Yellowstone's Specimen Ridge, where 27 petrified forests supposedly grew to maturity and were buried over millions of years. Clearly, such an arrangement can be created by a catastrophic event, calling into question the explanation for Yellowstone's petrified forests.

At the bottom of Spirit Lake, a 3-foot deep pile of bark has formed from the abrading of the logs floating in the lake. As the pile compresses, it is beginning to take on characteristics suggestive of coal formation. Formation of coal underwater is contrary to the peat bog model which has been the accepted theory. However, peat bogs forming today are always heavily interwoven with roots. Examination of coal beds reveals no roots but abundant flecks of bark (and the rare human artifact). Perhaps Mt. St. Helens will ultimately demonstrate that the uniformitarian coal formation theory needs to be superseded by a catastrophic one.

Perhaps in 500 years a naïve or poorly informed geologists will visit the Mt. St. Helens area and write a paper on the multi-million-year formation of the canyon, deposits, coal beds, and layers of petrified forests. Or is it possible that such a scenario is being played out today as geologists describe the Earth's features with a consistent uniformitarian viewpoint? Readers interested in learning more about Mt. St. Helens as it relates to catastrophic geology should view the video by Austin (1993).

The Geologic Column

The public, and even many scientists who are not geologists, have the opinion that there exists some physical entity known as the geologic column - that the strata around the world pretty well match the drawings for the geologic column in the textbooks. The truth is that nowhere is the entire geologic column represented, and at most places, more than half the column is missing. The geologic column is actually pieced together from segments at different places around the world.

In order to link the segments, similar strata have to be found and matched. That is, there has to be some overlap in the layers. However, correlations based on physical characteristics of the rocks are generally only useful over small regions of less than one hundred miles. For example, a gray shale sandwiched between a red sandstone on top and a white limestone beneath might be found at outcroppings a hundred miles apart, but one would not expect to find these same sediments throughout the world. To correlate strata worldwide, scientists use the evolutionary principle of faunal succession.

Faunal succession means that each period of Earth history had unique fauna that are represented by the fossils in the sediments laid down during the period. To simplify application of faunal succession, geologists and paleontologists have identified index fossils (see Chapter 2) that are used to correlate strata worldwide as being deposited at the same time. The evolutionary assumptions of uniformitarian deposition of the strata over four billion years and evolution of life over part of that time is the basis of fossil correlations and drives the interpretation of what geologists observe in the rocks.

The situation has developed in which the fossils determine the age of the rocks and the rocks establish the age of the life. Scientists would prefer to calibrate this relative system with radiometric dating. However, most rocks simply cannot be dated, and such dating is a slow and expensive process. As we will see in a subsequent section of this chapter, the fossiliferous age of the samples often guides the dating process. This entire unstable construct is shrouded in the fog of distant time, making the problems of evolution nearly impossible to see and the possibility of an Earth largely formed by catastrophism invisible to the naturalistic eye. Nevertheless, those not blinded by an erroneous presupposition have indicators that something is wrong.

Chapter 2 has already discussed the fact that the fossil record is not so clearly an unbroken succession of advancing life forms. But the lack of "a clear vector of progress" is further punctuated by anomalous fossils. Horse hoof prints and dinosaurs footprints have been found together. Human footprints and dinosaur footprints have been found together. Sometimes land animals, flying animals, and marine animals are fossilized side-by-side in the same rock. There are many places where fossilized trees, even entire forests, are found vertical, running through "millions of years" of strata. The strata containing these polystrate fossils could not possibly deposit in more than a few years. Coniferous pollen has been found in strata many millions of years older than the presumed first evolution of such complex plants.

Many fossils indicate very rapid burial. The fleshy jellyfish is often fossilized and thus had to be buried rapidly before decay. Mass burials of organisms in contorted positions are indicative of rapid burial. The abundance of fossilized land animals is a miracle based on common experience of rapid decay and scattering of the bones by weather and other animals. Something had to have buried these animals in a hurry. There are fossils of fish in the process of eating other fish, indicating a sudden, unexpected burial. We don't see fossils forming today, but there are vast fossil graveyards of dinosaurs swept into piles and buried.

Human artifacts are often found in ancient strata. Sandal prints have been found in several places in strata from 150 to 600 million years old. In one instance, the sandal stood on and compressed a tribolite, which supposedly became extinct 240 million years before humans evolved. A thimble, a spoon, an iron pot, a gold chain and many other such items have been found deeply buried in coal deposits and other rocks. (Brown 1995)

Geologists believe that the sediments were deposited either by wind or water, but that water was the principal factor. However, even the water-deposited sediments experienced long periods as dry land (by evolutionary reckoning), otherwise they would not contain terrestrial fossils. Almost all sediments throughout the world are flat, in parallel layers. Rarely does one find any evidence of ancient erosion or rock slides that would assuredly occurred during the time the sediments were above water. (The tilted or bent strata seen in roadcuts were obviously deposited flat and were deformed at a later time.) Many times where strata are missing from the geologic column, geologists blame these missing strata on erosion. However, the erosion always seems to be sheet erosion since the characteristics of the cutting erosion of gullies is usually not present. The near absence of true erosional features and the missing strata are evidence that the sedimentary rocks were deposited in rapid succession.

Meteors strike the Earth every day. In ancient times, these strikes may have been more frequent because the Earth and the planets have been sweeping the meteors from the solar system. However, meteorites are always found in the more recent sediments of just a few thousand years. This condition is consistent with a short period for deposition of the sedimentary rocks. Tektites are an Earth-origin glass material that has been blasted into space by meteor strikes and subsequently fallen back to the Earth. All tektites are found in the top few inches of the Earth's surface. Also, their delicate features have been preserved with little mechanical damage and etching from acid in the soil, indicating a young age. However, Australian tektites have been radioactively dated to be 700,000 years for one group and 4 million years for another (Barnes 1973). Why are they stratigraphically only a few thousand years old?

It takes more than a listing of anomalies or even a few mysteries to undermine a complete system of geologic thinking. Perhaps one day scientists will explain some of these anomalies. However, the issues raised in this section should cause the reader to question the authoritarian presentation of historical geology so often encountered in college courses and public television. The concept of a geologic column may well be a myth that is driving false interpretations of field observations. If scientists were to consider that the fossils may have been sorted by hydrodynamic principles and the local abundance of the extant species, rather than by time, a much different view of the Earth's history might be acceptable. For more information on this subject and for references for the anomalies described, the reader is directed to Brown (1995) and Sewell (1997). Creationist literature has numerous other examples.

The Grand Canyon

Anyone who has examined erosion on a freshly graded construction site has seen the gullies that can be carved by water in soft soil. Steep walls are formed as material is undermined and falls into the swiftly moving water. Branching gullies feed into the main erosional path. The appearance is amazingly like a view of the Grand Canyon from the air. Could the Grand Canyon also be the result of short-term erosion of soft sediments by a huge amount of water?

The Grand Canyon displays more strata in one location than any other place in the world. After noticing the majestic view, people begin to notice the strata. They are almost all perfectly parallel. If the sediments were deposited over hundreds of millions of years, in alternating marine and arid environments, would not one find more evidence of erosion and other topographical features? Was the land in northern Arizona perfectly flat and erosionless over hundreds of millions of years?

Or were the strata deposited in water over a relatively short period of time?

The conventional wisdom is that the Colorado River carved the Grand Canyon over many eons. If so, where did the sediments go? Where is the gigantic river delta? There is no evidence of a contemporary or ancient river delta. Perhaps these sediments were eroded by a torrent that deposited them over a wide area of California in a sheet instead of like river erosion. Scientists have recently begun to realize that meandering rivers like the Colorado just don't cut wide and deep channels. Some other process must have been responsible. The Grand Canyon offers many such challenges to uniformitarian geology. The following paragraphs of this section discuss only of few of the challenges to uniformitarianism. Interested readers should consult Austin (1994) for details.

The Redwall Limestone of the Grand Canyon is one of its most distinctive layers. Because limestone today is deposited very slowly, about one foot per thousand years, geologists believe that the Redwall Limestone is convincing evidence of the great age of the Canyon. However, a careful comparison of currently-forming lime muds and the Redwall Limestone show significant differences. Modern lime mud contain silt-sized crystals of the mineral aragonite and up to 10 percent calcite derived from skeletons of marine animals. On the contrary, ancient limestones, including the Redwall, are formed of clay-sized crystals of the mineral calcite (at 100 percent) with sand-sized skeletal fragments dispersed in the crystal matrix.

The geochemistry and fossils of the Redwall Limestone are suggestive of rapid precipitation from seawater, not the slow accumulation of skeletal material. For example, many fossils of marine mollusks known as nautiloids have been found in one place. Not only is the burial of so many a curious find, but they are oriented as if they were buried while attempting to maintain alignment in a swift current. Thus, like much of evolutionary biology and geology, the evidence is often more suggestive of catastrophism than uniformitarianism.

For many years, geologists have claimed that the Coconino Sandstone of the Grand Canyon was deposited by wind in a great desert - a condition inconsistent with a deluvian scenario. The evidence to support the arid deposition theory was the existence of footprints of four-footed animals moving across the sand dunes and the existence of the sand dunes themselves. However, rather than assuming the footprints were made in arid sand, Dr. Leonard Brand did careful experiments and other observations to show that the tracks are much more consistent with tracks made in water as the animals attempted to move uphill against a current. Similarly, the sand dunes have been shown to have been deposited underwater.

Most geologists believe that the shales of the Grand Canyon were deposited very slowly as clay accumulated in a river delta (although the delta cannot be found and the accompanying sand is absent from the shale). The evidence to support this view are: (1) thin laminae in the shale, (2) burrows of organisms within the shale, and (3) shrinkage cracks in the shale.

Until the Mt. St. Helens eruption, even creationist geologists believed that catastrophism would not produce very thin laminae. However, we now know that catastrophic processes do not necessarily homogenize the deposited material. Even storms have produced thin clay laminae, and laboratory experiments are showing the same effect. The burrows of marine organisms can be shown to be evidence that the sediments were not deposited over long time periods. In places where the organisms burrow, they disrupt the laminae leaving identification of individual laminae almost unrecognizable. The fact that the Grand Canyon shales have burrows but do not have disrupted laminae is indicative of a quick and deep burial that left no time for heavily burrowed laminae. Finally, both laboratory experiments and field observations have now shown that shrinkage cracks can form in clay sediments without desiccation, while deeply buried.

As we studied in Chapter 2, the fossils of the Grand Canyon do not show an orderly progression that would be predicted by biological evolution. Although some portions of the Canyon show progression consistent with evolution (perhaps this is a tautology - are they consistent with evolution because biology independently predicts it or do the fossils suggest the order of evolution?), the prevailing progression does not support evolution but a hydrodynamic ordering of fossils. "It is not clear whether the order of appearance of organisms in Grand Canyon, or anywhere on Earth, for that matter, is necessarily any different than a random order which a flood might produce." (Austin 1994)

Radioactive Dating

Radioactive decay is a process by which radioactive atoms spontaneously transmute to another lighter element with the emission of energetic particles. There are two major types of particles emitted: alpha particles and beta particles. Beta particles are light particles with the mass and negative charge of an electron. Their emission from the nucleus raises the positive charge (the atomic number) of the nucleus by one, but does not significantly change the mass (mass number) of the nucleus. In contrast, the alpha particle is relatively heavy, having two positively charged protons and two neutrons. Its emission decreases the atomic number of the parent atom by two and decreases the mass number by four.

For example, the most common isotope of uranium contains 92 protons and 146 neutrons for a total mass number of 238. Its symbol would be ^{238}U. It undergoes alpha decay producing the daughter product, thorium, with 90 protons and 142 neutrons for a total mass number of 234 (^{234}Th). Subsequently, ^{234}Th undergoes beta decay to protactinium, with 91 protons and 143 neutrons (^{234}Pa). After a long series of sequential radioactive emissions and transmutations, a final stable atom of lead is reached, with 82 protons and 124 neutrons for a total mass number of 206 (^{206}Pb). If one begins with a large number of uranium atoms, it takes 4.5 billion years for one-half of them to decay to lead.

The fact that there are established time periods for parent atoms to decay to daughter atoms raises the possibility of producing a clock based on radioactive decay. If one starts with one million pure ^{238}U atoms and allows them to undergo radioactive decay until there are 250 thousand ^{206}Pb atoms, then by the mathematics of radioactive decay (which we will not address here), approximately 1.9 billion years have transpired. Geochronologists use this principle to date rocks that contain radioactive atoms with parent-daughter relationships. Some important parent-daughter relationships used for radioactive dating are: ^{238}U-^{206}Pb (concurrent with ^{235}U-^{207}Pb), ^{87}Rb-^{87}Sr, ^{40}K-^{40}Ar. Another type of dating, based on carbon, ^{14}C, does not use parent-daughter relationship but the changing ratio of ^{14}C to natural carbon, ^{12}C.

The most direct application of radioactive dating is to igneous rocks and sometimes to metamorphic rocks. It is supposed that the geochemistry of the magma purified the parent atoms as they solidified into the lattice structure of the crystal grains. Direct dating of sedimentary rocks is not possible because the grains of sedimentary rocks would date the formation of the original igneous rock that had been weathered to form the sediment. Sedimentary rocks are indirectly dated by dating igneous layers that may lie above, below, or intruded into the sedimentary layer or by using index fossils to relate the layer to other sedimentary layers that have igneous rock timing relationships.

In order to judge the efficacy of radiometric dating, it is important to understand the assumptions behind the process. These are as follows:

1. *The radioactive half-life is known and has not changed during the entire time of existence of the rock.*

   Scientists use the term, decay constant, to specify the half-life. The decay constant is inversely proportional to the half-life; by specifying one, the other is specified. The assumption that the decay constant is indeed constant seems a logical one. However, radioactivity is a phenomenon that has only been known for about 100 years. Scientists cannot be assured that it is constant, and there are no fundamental principles demanding its invariance. If it were a slowly varying parameter (e.g., detectable change only over 50 years) then we would not know it today. Respected scientists, beginning with the great physicist P. A. M. Dirac, have considered the concept of the fundamental constants of nature being variable. Taylor (1991) presents a more complete discussion on the invariance of the decay constant.

2. *The abundance of the daughter atoms in the rock at the time of crystallization is known.*

   Geochronologists have devised several clever and plausible methods for dealing with this difficulty, including use of isotopic ratios with stable isotopes and the isochron method; however, the fact is, one cannot know what the original concentrations were.

3. *The rock dated has been a closed system throughout its entire history. That is, no parent or daughter atoms have entered the rock and neither have they escaped the rock.*

   This one assumption is probably responsible for more chronological discordances than any other difficulty in radioactive dating. Migration of either parent or daughter into or out of the sample can affect ages drastically. The problem is especially pronounced for potassium-argon dating because argon is a gas. Again, geochronologists have devised methods for accounting for this problem, but certainty can never be assured.

4. *The abundances of the parent and daughter atoms can be accurately measured.*

   In the early days of radioactive dating, laboratory measurements were an issue. Today, techniques such as accelerator mass spectrometry have nearly eliminated measurement error from the list of difficulties faced by geochronologists.

The popular press would have us all believe that radioactive dating gives amazingly precise and consistent ages. It is true that there is a fairly large body of data that is internally consistent. Many times superposition is demonstrated in the layers radiologically as well as physically. Dating of some samples by multiple methods often gives consistent ages. However, there is also a large body of dating information that defy logic. Such discordances are often blamed on detrital intrusions, leaching, hot water intrusion, inclusions of preexisting rocks, or a host of other reasons. Many times discordances just don't get reported at all because they do not match the uniformitarian hypothesis. The assumptions or measurement methods are sometimes adjusted until the "correct" answer is obtained.

Sewell (1997) reports a study by John Woodmorappe that involved 445 technical articles in 54 reputable science journals. In these reports, 350 radiometric dates are reported that conflict with fossil ages. Conflicting data are what make science exciting, so the existence of these discordances is not condemning. However, in almost every case, the fossil dates were given higher credibility than the radiogenic ones. Does the rejection of the radiogenic dates give confidence in what scientists think of their own methods? Woodmorappe cites one researcher who honestly stated:

> In general, dates in the "correct ball park" are assumed to be correct and are published, but those in disagreement with other data are seldom published nor are discrepancies fully explained.

It is difficult to argue against age determinations for objects that predate human history. However, measurements of items known to be recent (within human history) that show millions of years of radiometric age raise credibility problems. Submarine basalt from recent Kilauea volcano eruptions has dated at 22 million years. There are volcanoes on the rim of the Uinkaret Plateau of the Grand Canyon where the lava spilled over the rim. These lava flows are obviously the youngest in the

Canyon; however, they radioactively date as the oldest rocks in the Canyon. Although one K-Ar age is 10,000 years, several others using K-Ar, Rb-Sr, and Pb-Pb give ages from 117 million years to 1.32 billion years. Lava from the Mt. St. Helens eruption has recently been dated at 2.8 million years.

The case of skull KNM-ER 1470 is another interesting example. The skull was discovered by paleoanthropologist Richard Leakey (son of the famous Louis and Mary Leakey) in 1967. Although Leakey stated that the skull was virtually indistinguishable from modern man, it was found beneath a layer of volcanic tuff that was thought to be very old - too old to be over any human remains.

Samples of the tuff were sent to a laboratory for age analysis, but the result was 212 to 230 million years old. Because this result did not fit evolutionary theory, it was rejected. Over the next 10 years the rocks associated with the skull were dated many times, using various methods. Another laboratory dated some samples at 1.6 million years and others at 1.8 million years. Other measurements as low as 0.5 million years were rejected as anomalously young. Among the "acceptable" ages were clusters of results around 1.8, 2.4, and 2.6 million years. There was great debate on which age was correct. The geochronologists seemed to favor the 2.6 million year date. The anthropologists seemed to favor the 1.8 million year date. The debate was finally resolved (perhaps by exhaustion) at a compromise 1.9 million years. Again, the expected evolutionary answer is what guided the acceptability of the results.

Radiocarbon dating is used to date previously living objects such as bones (not fossils) and ashes from ancient fires. Cosmic rays striking the Earth's atmosphere cause the common isotope of nitrogen (^{14}N) to be converted to ^{14}C. The ^{14}C, which is radioactive, becomes integrated into the carbon cycle of the Earth, thus, becoming part of every living organism. When the organism dies, ^{14}C is no longer being maintained through the carbon cycle but begins to decay away with a half-life of 5,730 years. Therefore, the $^{14}C/^{12}C$ ratio decreases over time, giving an indication of the age since death.

The method is based on the assumption that the $^{14}C/^{12}C$ ratio has been known and constant since the organism was alive. The inventor of the method, Willard F. Libby, measured the ^{14}C production and decay rates and concluded that the production rate was greater than the decay rate, indicating that the Earth had not existed long enough for equilibrium to be reached. Since the half-life is 5,730 years, a few half-lives or perhaps 30,000 years would be needed to reach equilibrium. Libby decided that disequilibrium was not plausible because of the great age of the Earth and attributed the discrepancy to experimental error.

However, more recent studies have shown that ^{14}C really has not yet reached equilibrium. This would mean that all objects dated under the equilibrium assumption would have to be corrected downward in age. The method is only good for about 10 half-lives (perhaps 60,000 years). Corrected dates of formerly living entities would have to be adjusted downward, perhaps as much as 50,000 years for the oldest dates where the disequilibrium was most pronounced. This places all the dates within approximately 10,000 years.

When ^{14}C dating was first invented in 1947, the scientists performing the measurements were physicists who generally do not have a stake in evolutionary theory. They tended to date everything they could get their hands on, even if the specimens were supposedly millions of years old. Dinosaurs, coal, Neanderthal Man, and many such specimens were dated as only a few thousand years old.

In fact, all formerly living specimens produced dates within the 50,000-year window of the method's limit at the time. Not fully realizing or caring about the political ramifications of these ages, the physicists continued to report them until it became unacceptable to do so.

Given the difficulty in verifying the assumptions underlying radioactive dating, the trial-and-error approach to achieving politically acceptable results, and the vast numbers of inconsistent and incongruent results, one should accept radiometric ages with caution. Scientists should be foremost in warning the public about the tentative nature of radiometric dates. Unfortunately, they are foremost in using radioactive dating in the propaganda war promoting the world view of naturalism and its pet theory, biological evolution.

Radiohalos

As discussed in the preceding section, the utility of radioactive dating is dependent on the assumption that the radioactive half-lives of the various radionuclides have always been constant. At one time, radii of radiohalos were thought to provide evidence of the invariance of radioactive decay rates. In the early 1960's, the question of whether half-lives were constant led physicist, Robert Gentry, to begin investigating these halos.

Radiohalos are microscopic spherical discolorations in rocks resulting from emission of alpha particles from radionuclides that are concentrated in tiny grains (typically about one micron diameter) embedded in the host rock. As the alpha particles are emitted in three dimensions, they ionize the atoms of the rock's crystal structure, producing a darkening of the rock. An example is a ^{238}U halo in which there are eight concentric halos resulting from the eight alpha emitters in the uranium decay series.[34]

Although Gentry eventually demonstrated that there is no useful correlation between measurements of radiohalo radii and variations in radioactive decay rates, he discovered something much more important: certain radiohalos may be the imprimatur of an instantaneous creation *ex nihilo*. These important radiohalos are the product of polonium decay. There are three types of polonium radiohalos: ^{218}Po, ^{214}Po, and ^{210}Po, with half-lives of approximately 3 minutes, 164 microseconds, and 138 days, respectively. The reason these facts are significant needs some explanation.

According to the current theory for the origin of the elements, nuclear reactions in the centers of stars are responsible for the nucleosynthesis of elements beyond hydrogen and helium. When these stars undergo supernovae, they disperse the heavier elements throughout the universe. As new stars and solar systems are born, these elements become incorporated into the matter of planets. The newly formed planets are very hot and must solidify and cool down for life to evolve and do such things as discovering radiohalos.

Under the currently accepted theories, if the polonium that was in the halo centers were primordial, that is, the direct result of nucleosynthesis, it would have to have survived radioactive decay from the time of its synthesis until the rocks cooled enough to record the image of a halo (hot rocks do not record halos). Such a process could take billions of years. Because the polonium half-lives are so short, such a hypothesis is unreasonable. However, if instantaneous creation of the Earth

---

[34] Only five halos are actually observed because of the similarity in energies of some of the alpha-emitting radionuclides.

occurred, then the polonium and the solid rock that contains it could have been created together, recording a halo.

The other possible explanation is that the polonium is not primordial. All three polonium isotopes are natural decay products of ^{238}U. Uranium has a half-life of 4.5 billion years - long enough to survive the formation of the solar system and produce polonium on a cool Earth. However, Gentry has done meticulous experiments to demonstrate that the polonium is not radiogenic (that is derived from uranium decay).

Because of its implications for the origin of the universe and the life within it, Gentry's work has attracted the attention of world-class scientists. Although the majority opinion is clearly on the side of non-primordial polonium, no non-primordial hypothesis for polonium radiohalos has been successfully proposed and defended. The arguments against primordial polonium are generally hand-waving arguments without theoretical or experimental support.

Gentry's work is strongly suggestive that either the rocks were created instantaneously or the radioactive decay rate was different in the past than what it is today. A full account of Gentry's work and his experiences in publishing it are reported by Gentry (1992).

Conclusion

Biological evolution has demanded that the Earth give it time to produce the impossible - life from non-life, the diversification of all the species that have ever lived from a single proto-cell. Geologists have been forthcoming with that time. However, much of the evidence for an ancient Earth can be reinterpreted within a catastrophic framework.

Once we yield the point that a Creator was necessary for life, is there any reason to insist that the Earth has to be old? Without the bias of naturalism, the Earth can be seen as young. Scientists should study the Earth to see what scientific treasures it will yield - not to support a philosophical worldview. Even the intelligent design theorists would do well to take the lessons they have learned from exposing the façade of biological evolution and see if there is a similar lesson to be learned about uniformitarian geology.

Seven

Richard Owen's Fossils

Special Challenges for Creationists

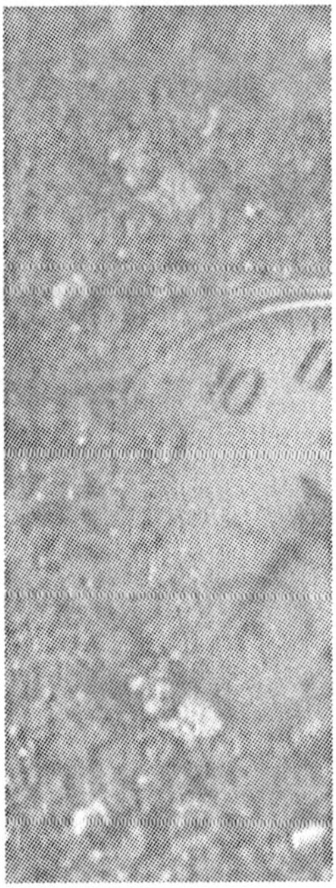

Richard Owen (1804 - 1892) was the world's leading nineteenth century anatomist and one of the first-ever paleontologists. He was also the founder of the British Natural History Museum. Although known for first defining the term "homology" (a concept that was discussed in Chapter 2), Owen's most famous achievement occurred in 1842 with the identification of a whole new order of reptiles that he named *Dinosauria*, a word he derived from the Greek, *deinos* (terrible) and *sauros* (lizard).

Richard Owen was also the discoverer of Eohippus. Although modern evolutionists believe Eohippus is the dawn horse (first horse-like ancestor), in 1841, Owen did not make any connection of this animal to a horse. He believed it to be related to the modern hyrax (rock badger). Thus Owen named it *Hyracotherium*, its more scientific name that is still used today. In 1863, Owen reported the first specimen of an unusual bird, *Archaeopteryx lithographica*, which has played prominently in the case for evolution.

Given these major achievements in areas of critical interest to evolution, one would think that Owen would be a staunch Darwinist. However, Richard Owen was an outspoken critic of Lyell's geology and Darwin's descent with modification - not for religious reasons but for scientific ones. Therefore, in an 1855 address to the Royal Institution of Great Britain (four years before *Origin*), Owen highlighted the differences in the anatomies of apes and humans. He stated that these differences, which included the prominent brow ridge on the apes, would preclude any evolution of man from ape. It must have been quite a surprise to Owen when one year later, quarrymen discovered bones in a cave of the Neander Valley in Germany. The bones, which were obviously human, included a skull with a prominent brow ridge. After Darwin published *Origin* in 1859, the so-called Neanderthal Man was used as striking evidence for the evolution from ape to man.

This chapter could be designated as Richard Owen's chapter. It discusses human fossils, dinosaurs, *Archaeopteryx*, and the horse fossils - subjects of special interest to Owen. Because of their seemingly compelling argument for evolution, these subjects are worthy of special attention.

Human Fossils

An interest in the origin of humanity seems to be hard-wired into the psyches of all people. Every culture, both current and past, has had an origins myth. The origins story for many Christians, Jews, and Muslims is that which is recorded in the book of Genesis in the Bible. The origins story of naturalism is the purely mechanistic view of humanity evolved from the ever expanding flow of life emanating from the original proto-cell.

The intensity of fascination with the subject has made it difficult for scientists to objectively study human origins. Perhaps recognizing the unusual degree of strife and emotion among paleoanthropologists, evolutionist Douglas Futuyma writes (Futuyma 1998):

> However, individual scientists are often far from objective. Nowhere is this more true than in the area of human evolution and variation. Because of their social and philosophical implications, these subjects stir up more than the usual amount of passion in scientists, as in everyone else.

Human evolution is likewise an emotional topic for creationists, "the topic that ignites the creationist's fulminations against evolution," according to Futuyma. Evolution of humans would explain the human condition without reference to sin and would obviate the need for a Savior. It is a most direct challenge to the biblical account of beginnings. Having trained for the ministry, Darwin was well aware of the outrage that would ensue if he proposed human evolution. Thus, he limited its discussion in *Origin* to the following on the next to last page (Darwin 1859): "Light will be thrown on the origins of man and his history."

Creationists have to confront the evidence for human evolution. There are, after all, those hominid fossils that can be arranged in an order from ape-like to human-like. Chapter 2 challenged the idea that the fossil record reveals transitional forms, but it deferred discussion on hominid fossils. Given the importance of human evolution and the challenge the hominid fossils offer to creationists, the subject of human evolution has been reserved for this current chapter.

Hominids are species of the family *Hominidae* (order Primates). Only one species is extant today, *Homo sapiens* or human beings. However, paleoanthropologists classify several extinct species as hominids, including several species of the genus *Australopithicus* and several others of the genus *Homo*.

The family of Primates most resembling *Hominidae* is *Pongidae*, which includes the gorilla, the chimpanzee, and the orangutan (but not gibbons, old world monkeys, new world primates, or lemurs). *Pongidae* and *Hominidae* are believed by many to have evolved from a common ancestor during the late Miocene epoch, 11.2 to 5.3 million radiometric years ago. To clearly establish the phylogenetic history of humans, scientists would ideally seek to find transitional forms from both modern humans and Pongidae (the great apes) that converge on the common ancestor. However, evolutionists recognize that the fossil record does not contain any precursors to modern apes.

On the hominid side, there are fossils that scientists have classified as primitive to modern man. However, with a few exceptions, the fossils are primarily teeth and skull fragments, and many of these are not sufficiently preserved to unambiguously distinguish one species from another. To make matters more difficult, the absence of fossils from the ape-human common ancestor and all forms leading to modern apes means that scientists are forced to use modern ape characteristics as surrogates for those of the common ancestor. Thus, primitive hominid fossil morphology is categorized in an evolutionary context between modern apes and modern humans, not the common ancestor and modern humans. Such an extrapolation may not be warranted.

Some of the characteristics that paleontologists use to distinguish ape-like characteristics from human-like characteristics are provided in the table below. Most of the characteristics have to do with the cranium. The paucity of post-cranial fossils makes determination of the mode of locomotion and posture difficult, feeding substantial debate among proponents of rival theories. There are many more characteristics used by scientists; the following table is a gross simplification.

Hominid Species

Australopithecines are the oldest of the fossil hominids. There are four species of this genus that are generally accepted within the paleoanthropological community. These are *afarensis*, *robustus*, *boisei*, and *africanus*. The earliest of these species is *afarensis*, which includes the famous "Lucy," discovered by Donald Johanson in 1974. Some scientists believe there may have been four separate lineages developing from *afarensis*, with *robustus* and *boisei* separately branching in one direction to an evolutionary dead end and *africanus* and the genus, *Homo*, separately branching in another direction. The phylogenetic relationship is much in question.

Afarensis is very "primitive" in evolutionary terms, meaning that it has characteristics close to those believed to be held by the ape-human common ancestor of over 5 million radiometric years ago. These features are a protruding jaw (prognathism), a flat cranial base,

Characteristic	Human-like	Ape-like
Skull volume and shape	Larger and rounded	Smaller and less rounded
Orbital brow ridge and groove	Absent	Present
Eye orbits	Larger with more gracile eye rings	Smaller with more robust bony rings
Protruding muzzle	Absent	Present
Foramen magnum placement (exit opening from skull to spinal chord)	Centrally located	Located more posterior
Palate and mandible shape and size	Parabolic and smaller	Rectangular and larger
Tooth size and shape	Smaller	Larger
Extremity bone size and shape	More straight with smaller upper limbs and larger lower limbs	More curved with larger upper limbs and smaller lower limbs
Pelvis shape	Adapted for upright bipedality	Adapted for quadruped locomotion

large canine teeth, long arms relative to the legs, a small cranial capacity (about 400 cc), and curved bones in the fingers and toes. However, *afarinsis* is believed to have the capacity to walk upright. *Robustus* and *boisei* had similar characteristics but are found in sediments of later radiometric age.

Africanus is the most gracile of the australopithecines and, according to evolutionists, may or may not lie in the lineage to *Homo*. Whether considered directly or indirectly a link to humans, *africanus* is considered very important in the study of human origins. It has a slightly larger cranial capacity (450 cc) and the skull is obviously transitional in appearance. It is thought to have lived 3.0 to 2.0 million radiometric years ago. There is no evidence that *africanus* used tools.

The fossil evidence for australopithecines is not large, but it does show a slightly increasing cranial volume with decreasing radiometric age (whether actual or an artifact of preconceived ideas is not known). Fossils of perhaps 100 separate individuals have been discovered, all in Africa. It should be noted that most "individuals" are only fragments of a single bone. The most complete individual is Lucy, which contains a 40 per cent complete skeleton but no skull, hands, or feet. Thirteen "individuals" were found with Lucy, but none as complete as she; none had skulls.

Paleoanthropologists are in general agreement that three *Homo* species have existed: *habilis, erectus*, and *sapiens*. *Habilis* is the earliest and thus most difficult to define. Some scientists believe that *habilis* is just another australopithecine. With only a handful of fossil "individuals" to support the species, some uncertainty is understandable. The brain capacity of *habilis* is larger than that of the australopithecines (610 to 800 cc), but it is thought to have coexisted with them. The *habilis* limbs had ape-like proportions, but the skull had less prognathism. Crude tools have been found near *habilis* finds, so as its name implies (handy man), *habilis* is thought to have used tools.

Erectus has many modern human features. The cranial capacity is around 1,000 cc. (Modern human cranial capacity ranges from 850 cc to 1700 cc, with an average around 1400 cc.) The skull is rounded with a much less prognathous face. *Erectus* is clearly associated with stone tools, and use of fire was widespread. *Erectus* has been identified in Africa and Asia in sediments from about 1.6 million to 27,000 radiometric years ago. Less than 100 "individuals" have been found.

Sapiens fossils, or modern man, have been identified in sediments as early as 300,000 radiometric years ago, indicating co-existence with *erectus*. Some older fossils are known as archaic *sapiens*, but differ from modern *sapiens* in only minor characteristics. These archaic forms include the well known Neanderthal Man and Cro-Magnon Man. Although Neanderthal Man had dense bones and a projecting brow, they had average cranial capacities somewhat larger (1500 cc) than modern man. A fleshed and dressed Neanderthal would fit into the norm of modern human appearance.

Famous Fossil Men

We have all seen representations of ancient man as stooped, covered in hair, naked, cave dwelling, and somewhat ape-like in appearance. Even Neanderthal Man and Cro-Magnon Man, who are fully human [Homo sapiens (archaic)], have been presented in this fashion in books and museum exhibits. What were these men really like?

The Neanderthal Man (Neandertal is the most recent spelling) may not have been a cave dweller, but clearly used caves for burial and preservation of artifacts. Over 60 individuals of Neanderthal characteristics have been found, mostly in Europe but also in Africa and Asia. These people were artists and practiced ritual burial of their dead, apparently no more primitive than some modern cultures. Some studies have suggested that the curved leg bones and brow ridge are the result of a vitamin D deficiency, acromegaly, rickets, and arthritis, making them truly modern Homo sapiens after all. One Neanderthal skeleton was found buried with a suit of chain armor. One living Neanderthal was discovered in the Philippines in 1910. A single Neanderthal skull has been found in California where it does not belong.

Cro-Magnon Man was also an artist, producing well developed paintings of themselves as well as ivory carvings. One picture shows a mother and child in dresses and hairstyles. The brain capacity of Cro-Magnon is even larger than the Neanderthal. Their bodies are larger than modern man and well proportioned. Perhaps man is degenerating from his Cro-Magnon days rather than evolving. Cro-Magnons used tombstones to mark the burial of their dead.

Java Man is represented by so few bones and is clouded in such controversy that it essentially provides no evidence at all for human evolution. Peking Man, which started with a single tooth, is primarily represented by a large number of skulls found in the same location. Every skull is battered and the base is broken in a peculiar fashion. This suggests that the skull was decapitated and the brains withdrawn as is done in some cultures today for food. The animals that were eaten may have been an extinct primate. A furnace operation was apparently associated with these skulls. Peking Man does not offer any compelling evidence for human evolution.

Interpreting the Evidence

Clearly, animals and humans different than those that exist today are represented in the fossils. The question is whether these animals and humans represent a progression of evolution. It is possible to arrange the fossils in a sequence from ape-like to human-like, when using onc or two gross characteristics. However, careful examination reveals that the progression is more mosaic. When one characteristic is used, the fossils arrange in one order and when another characteristic is used, the fossils arrange in another order. This is strong evidence against an evolutionary origin of the represented species.

The mosaic nature of the species could be explained by the low number of individuals represented and the lack of complete or even partial skeletons for the individuals. However, this argument is one in favor of a non-evolutionary interpretation. There is huge variation among modern humans, ranging in size alone from the Pygmy to the late wrestler Andre the Giant. There are ethnic groups today that have a pronounced eyebrow ridge. The same diversity in morphology would be expected to be found in the great apes and their "evolutionary" precursors. Without more complete knowledge of the characteristics of a species, including the degree of statistical variation within the species, it is very difficult to organize fossil finds of a few bone fragments into species and place the species in a phylogenetic order. The hominid fossil record is just too sparse to draw any firm conclusions in favor of human evolution.

Unfortunately for good science, the primary goal of paleoanthropology is to find evidence to support a preconceived theory. Couple this nonscientific practice with the pressure to produce missing links that encourage funding, and we have a recipe for fanciful speculation. In this environment, discoveries contrary to the accepted presuppositions are often discarded as anomalies.

One example of a largely ignored discovery is the Lady from Guadeloupe. In 1812, a fully human female skeleton without feet or head was discovered in hard limestone believed to be 28 million years old. The twisted but still articulate fossil was on display in the British Museum for over a century. This discovery does not fit the evolutionary framework for two reasons. First, it is too old to be placed into the hominid phylogenetic tree. Secondly, no hominids were supposed to be in the Western Hemisphere. One wonders how many modern human fossils have been overlooked and how many degenerate or diseased ones have been identified as ancestral.

When hominid fossils are found that do not match current norms for human anatomy, scientists have a great tendency to assign the finds into an evolutionary framework. This is not fraud but lack of objectivity. However, we already know from the fossil record that today's families (in the taxonomic sense) contain extinct species and that the great apes look more like humans than any other animals. How close to human appearance can extinct *Hominidae* species get before creationists start to get nervous? Could extinct primates look more like humans than today's apes and still not represent links in an evolutionary chain? Are there extinct primates, which may look somewhat human-like, incorrectly being portrayed as human precursors?

Some creationist books are very settled in the matter. Australopithecines and *Homo habilis* are extinct primates and *Homo erectus* and *Homo sapiens* (archaic) are fully human (but a little different from today's norms). This seems to be a logical conclusion for one who is prejudiced towards divine creation and not towards naturalistic origins. However, the evidence from which one judges the matter is just as scarce for the creationists as it is for the evolutionists. As with some of the other arguments for or against evolution, one's presuppositions seem to be the deciding factor.

Given the extreme improbability of life originating from non-life (see Chapter 3), the lack of a mechanism to evolve one species into another, the lack of evidence that macroevolution has even occurred (see Chapter 2), there is very little to compel an objective observer to embrace human evolution. Although the hominid fossil evidence does little to support a creationist view, neither does it provide coercive evidence for evolution. The fossils' superficial suggestion of human evolution, however, has been used to further the worldview of naturalism. Those that do not embrace a purely mechanistic understanding of life can find nothing in the hominid fossil record to pressure them into a different opinion.

Dinosaurs

Before Richard Owen identified dinosaurs as a distinct type of creature, scientists had little concept of the life forms that produced those large bones that, since 1818, were being found in England and North America. However, since 1842, dinosaurs have generated great scientific and public interest. The intensive search for more dinosaur specimens has resulted in more than 500 genera and 1,000 species. The abundant fossil remains have been found on every continent on Earth.

Dinosaur Facts

Scientifically speaking, the term, dinosaur, applies only to two orders, the *Ornithischia* (with bird-like pelvis) and the *Saurischia* (with a reptile-like pelvis). The *Ornithischia* are thought to have been plant eaters. Some of the best known representatives are the famous iguanodon (of Disney's *Dinosaur* fame), hadrosaurs (duck-billed dinosaurs), and ceratopians (including triceratops).

The *Saurischia* are comprised of sauropods, which were large herbivorous dinosaurs with long necks and tails and bulky legs and theropods, the most fearsome bipedal carnivores. The sauropods are represented by the huge brontosaurus (more correctly termed apatosaurus). The therapods are represented by the notable *Tyrannosaurus rex* featured in the film, *Jurassic Park*.

Besides the true dinosaurs, several other reptilian creatures are often classed as dinosaurs in common usage. These include the giant winged pterodactyls; the long-necked plesiosaurs, which lived in the sea; and the ichthyosaurus, a dolphin-like reptile.

Paleontologists believe that dinosaurs dominated the Earth during the Mesozoic Era, which included the Triassic period from 251 to 206 million radiometric years ago, the Jurassic period from 206 to 144 million radiometric years, and the Cretaceous period from 144 to 65 million radiometric years ago. At the boundary between the Cretaceous and the Tertiary (Cenozoic Era) periods, the so-called K-T boundary, dinosaurs suddenly no longer appear in the fossil record. Although evolutionists are not fully unified on either the cause or the abruptness of the extinction, the dominant extinction theory is a catastrophic one: large meteor impact. Of course creationists believe the extinction was due to the great flood of Genesis.

The Challenge to Creationists

Most of the facts about dinosaur fossils do not fit well with the perceived creationist views of Earth history. From a biblical viewpoint, why would God create such fierce creatures? Why were they so large and abundant in variety? Could they possibly fit onto Noah's Ark? Why are there no descendants of dinosaurs today? Such questions can be challenging to individuals inclined to believe in creationism but not yet convinced of its truth.

Answers to these questions are theological rather than scientific.

So far, this book has avoided theological arguments in favor of descriptive science, combined with logic (and a sprinkling of philosophy). The fact that we now turn to theology is not an admission of scientific defeat and not a "God of the gaps" argument. Other than the assumed great age of the dinosaur fossils, creationists do not contest the facts of dinosaurs. Therefore, there is no need for a scientific argument (Chapter 6 addresses the dating issue.). However, creationists must demonstrate that the existence of dinosaurs in Earth's history does not falsify the creationists' position.

Theological Arguments (in a nutshell)

Probably all creationists would believe that only God can create life. Thus, it follows that God created the dinosaurs. However, what God created and declared "good" is not necessarily the same life as we see in the world today. God created a paradise in which there was no death. As a result of mankind's sin, the Earth and the life it contained was changed. *"[C]ursed is the ground for thy sake ... thorns and thistles shall it bring for to thee"* Genesis 3:17, 18. Trials and death were inserted into the world to provide a new kind of education for mankind, in hope of ultimately redeeming the creation.

Sin degraded the behavior of humanity and all life. A carnivorous food chain developed. The second generation of humans to have ever lived produced one who murdered his brother. After approximately 2,000 years, *"God saw that the wickedness of man was very great in the earth ... and the Lord said, I will destroy man whom I have created from the face of the earth; both man, and beast ...."* Creation degraded so rapidly that God decided to get a nearly fresh start. He preserved eight individuals and selected animals in an ark and destroyed the world with a flood.

Today we have fierce carnivorous animals such as lions. When God redeems the creation, it will return to the condition in which it was created, "*The wolf and the lamb shall feed together, and the lion shall eat straw like the bullock*" Isaiah 65:25. Apparently, the lion was not originally carnivorous. Today we have other creatures so fierce that the lion pales in significance. The malaria virus or the Ebola virus are more vicious predators than even the tyrannosaurus. Did these fearsome life forms have a gentle and useful precursor in the Garden of Eden? Perhaps the perversion of nature can be explained by the New Testament concept, "*An enemy hath done this,*" Matthew 13:28.

We do not know what creatures were originally created. Whatever the precursors were to dinosaurs, they were not what we view them to be today. Sin has corrupted a world including the originally created life upon it. In fact some things were so corrupted that God destroyed them altogether with the Genesis flood. Perhaps dinosaurs were sufficiently perverted from God's design that he eliminated them, and scientists now see the result in the so-called Mesozoic Era. Although there is no biblical proof of this history for the dinosaurs[35], the abundance of dinosaur fossils and the mass extinction at the K-T boundary is very much consistent with the biblical description of the world's history. Noah's Ark would not have needed to contain any dinosaurs.

---

[35] Some creationists believe that the behemoth described in Job 40 may have been a sauropod. Similarly, the leviathan of Job 41 may have been a pliosaur or an ichthyosaurus. There is neither biblical nor scientific evidence to support this interesting speculation.

Dinosaur Uncertainties

If dinosaurs lived concurrent with mankind as creationists believe, one would hope to find evidence of coincident human and dinosaur fossils. Such a find would be devastating to the concept of the geologic column (two orders of magnitude difference in age). The most famous examples of human footprints with dinosaur footprints are those in the limestone of the Paluxy River near Glen Rose, Texas. Some creationist scientists are convinced that the evidence does indeed suggest humans and dinosaurs walking in the same soft mud. Other creationist scientists disagree. Given this uncertainty, creationists should use caution in citing the Paluxy River footprints in support of a young Earth.

For decades, there has been some small uncertainty among evolutionary scientists about whether the dinosaurs really were reptiles. There are a few cases, such as Stegosaurus, where the animal is clearly armored, but there are and were armored mammals such as ten-foot long, extinct glyptodont (a giant armadillo). In most cases, we simply don't know anything about their external appearance, whether they were scaled or covered in fur. However, a very recent discovery of a fossilized tescelosaurus (a plant-eating sauropod of 13 feet long) apparently had its heart fossilized as well. "This challenges some of the most fundamental theories about how and when dinosaurs evolved," states Dale Russell, one of the scientists involved in the research (CNN 2000).

Most reptiles have a three-chambered heart. Some, like crocodiles, have a four-chambered heart but a double aorta. The tescelosaurus was found with a four-chambered heart and a single aorta, just like mammals. Biologists have believed that the three-chambered heart is associated with cold-blooded animals. Although this is only a single dinosaur among many, the discovery raises the question of whether dinosaurs might be warm-blooded mammals[36]. It also strongly suggests that fossilization can be a very rapid process.

Fossilization, in which molecules of the bone are replaced one at a time by rock mineral molecules, is called permineralization. One might think that it would take many millions of years to occur. Yet there are a few fossils of soft body parts, like the dinosaur heart, that should have decayed in a matter of weeks. However, experts in the subject state that modern bones that fall into mineral springs can become permineralized within a matter of weeks (Wieland 1998). Without permineralization, bones would disintegrate over hundreds of years unless they were preserved from bacteria and water.

There are a few cases where dinosaur bones have been found preserved in some manner such as in ironstone modules. "The nodules prevented water from invading the bones, which for all intents and purposes, cannot be distinguished from modern bone" (Wieland 1998). For bones to be preserved 65 to 200 million years without fossilization is amazing - perhaps non-credible. Some dinosaur bones still contain red blood cells and hemoglobin. For these complex and fragile organic constituents to remain for more than a few thousand years is even more unlikely. One must question whether dinosaur fossils are as old as indicated by the geologic column dates.

[36] Given how recent this discovery is, scientists not directly involved should be cautious about drawing conclusions until the discovery is substantiated by other investigators.

Creationists do not have especially satisfying scientific answers for the existence of so many dinosaur fossils. Nevertheless, dinosaurs do not falsify the creationist viewpoint. Given the many difficulties of evolutionary science, scientists with minds open to a creationist view of life need not be influenced towards evolution by the dinosaur evidence. Indeed, the uncertainties associated with the geologic column and the dinosaur fossil evidence clearly indicate that more information is needed before drawing many conclusions about Earth's history from such fossils.

Archaeopteryx

Archaeopteryx is a rare bird. It is one of the most important fossils ever discovered and represents the best hope of a transitional form that evolutionists have ever encountered. The first *Archaeopteryx* was found in 1861; however, there have been very few others discovered, with most of them having been found in Germany. The fossils may represent two species.

The skeleton has many reptilian features, leading some evolutionists to believe it is actually a theropod (bipedal dinosaurs with some avian characteristics). However, *Archaeopteryx* has the one defining characteristic of birds - feathers. The presence of feathers places it firmly in the bird classification. This bird had teeth, however, something that no modern birds have. Some scientists believe it could fly; others believe it could not. Whether this creature is a dinosaur, a bird, a hybrid, or a dinosaur becoming a bird has been debated among evolutionists and creationists alike.

At one time, creationists proclaimed that the feathers on *Archaeopteryx* were a hoax. This claim was based on a series of articles by Drs. Fred Hoyle, N. Wickramasinghe, L. Spetner, and others, which stated that chicken feathers had been pressed into a grout that had been spread on the fossil. The British Museum of Natural History has refuted the idea of a hoax. Some creationists today are more careful about challenging the authenticity of *Archaeopteryx*.

Textbooks have proclaimed *Archaeopteryx* as convincing evidence of the truth of evolution, that it is truly a transitional form. Given how little we know about this creature today, all scientists, whether evolutionists or creationists, should resist dogmatism on the subject. Creationists should be willing to acknowledge the obvious: *Archaeopteryx* has both reptilian and bird features. The creationists would say that this is a mosaic, like the duck-billed platypus. The evolutionists would say that it is a transitional form. Nevertheless, the absence of other transitional forms in the fossil record is only highlighted by the presence of the one example that looks a lot like a transition, whether it is or not.

In conclusion, those examining the fossil record to learn the truth (rather than to establish a preconceived idea) should not find in the *Archaeopteryx* coercive evidence for evolution. However, it is clear that *Archaeopteryx* is something that evolutionists can point to as evidence in their favor. Given the lack of other evidence for evolution, *Archaeopteryx* offers no challenge to the creationist understanding of origins.

Horse fossils

The phylogeny of the horse is believed to be one of the best documented in paleontology. Older textbooks show graphics of four or five horse-like skeletons or reconstructions arrayed in a row - starting with the smallest and most primitive and ending in the modern horse. Today, evolutionists believe that the horse sequence is not nearly so linear - that the lineage is filled with branches and dead ends. Approximately 30 genera are known. However, the arrangement of the branching phylogeny varies greatly from author to author. The textbook graphics generally show representatives of four genera: *Hyracotherium*, *Mesohippus*, *Merychipus*, and *Equus*.

The so-called Eohippus (literally dawn horse) or *Hyracotherium* is thought to be near the common ancestor of all odd-toed hoofed mammals (the order *Perissodactyla*), which includes horses, asses, zebras (the Equines), rhinoceroses, and tapirs. This animal had four toes on the front feet and three on the hind feet. The toes were nearly vertical, causing the *Hyracotherium* to walk on tiptoes. Each toe ended in a small hoof. The dentition suggested a browsing diet (leaves of trees and bushes). Specimens range in height from 1 to 2 feet. The *Hyracotherium* was so unhorse-like that its connection to equines was not apparent until evolutionists thought that it might be the dawn horse.

The *Mesohippus* was a three-toed animal the size of a collie dog. It was very horse-like in appearance. Its dentition was still suggestive of browsing. The next genus, *Merychippus*, had teeth much like the modern *Equus*, i.e., grazing dentition. It was the size of a modern pony. Although *Merychippus* had three toes, in many specimens, the center toe is much enlarged and had a well-developed hoof.

The modern horse belongs to the genus, *Equus*. There is amazing variety of forms and sizes among *Equus*. In just the horses alone the size can range from about 20 inches to over 6 feet. Modern horses can have from 17 to 19 pairs of ribs. Although *Equus* generally has only one hoofed toe, there are some horses in the American southwest with three toes of almost equal size. Presumably, if selected skeletons of modern Equus were placed into the fossil record at the appropriate places, it would be very tempting for the unsuspecting evolutionist to arrange them in a phylogenetic order.

To the objective observer, the fossil horse sequence does not present an example of linear or even branching progression from primitive to modern. It is an illustration of the abundance of genetic depth contained in a created kind. The *Hyracotherium* is most likely just what Owen said it was - a type of rock badger. Thus the anchor to this sequence does not even exist. The other 25 or so genera are examples of horse-like creatures that, although somewhat different than those of today, fit within the bounds of modern *Equidae*.

Conclusion

Discussions in text books and the press about human fossils, dinosaurs, *Archaeopteryx*, and horse fossils over the last 100 years have convinced many scientists and other intellectuals that the evolutionary framework for life's origins is an impregnable fortress of intellectual solidarity. As we have seen over and over, however, evolution has been built preconception upon preconception such that it only <u>appears</u> to be that way. The seemingly very complex but stable collection of interconnected theories and facts that is proclaimed by evolution's apologists is not so convincing when viewed without the bias of naturalism.

The fact that intelligent scientists can examine the same evidence and arrive at such diverse conclusions as special creation or evolution, raises the question of whether science can ever arrive at truth. Is scientific truth achievable? Must the historical sciences always remain in a fog of uncertainty? These questions are the subject of the next chapter.

Eight

How Do We Know Truth?

Over the millennia of human intellectual activity, there is one philosophical question that has consumed more brain energy than any other. Perhaps the most famous phrasing of the question was by a Roman governor of Judea, Pontius Pilatus. When Jesus was brought to him by the Jews for sentencing, he questioned Jesus briefly, asking, "*Art thou the King of the Jews? ... what hast thou done?*" Jesus answered, "*To this end was I born, and for this cause came I unto the world, that I should bear witness unto the truth. Every one that is of the truth heareth my voice.*" Pilate then countered with a question that still waits for an answer, "*What is truth?*" John 18:33-38

This question constitutes one of the four main fields of philosophy and has been addressed by nearly every great philosopher - epistemology. (The others are metaphysics, logic, and ethics.) Epistemology is the study of the origin, nature, limits, and methods of knowledge, or simply, what is truth and how does one obtain it?

Development of the Scientific Epistemology

Prior to the time of Francis Bacon, the intellectual worldview included God reigning over the universe through his prime creatures, humans. Divine revelation was the final arbiter of truth. Unfortunately, after more than a thousand years of church-dominated society, in which ignorance was maintained among the general population, superstition was prevalent. Knowledge was authority based. The concept of individuals seeking and gaining knowledge through observation and experimentation was unheard of. The superstition and ignorance of the period makes it difficult for us near the end of the Modern Age to visualize this worldview in a positive context.

However, the independent thinking that began in the Renaissance and blossomed in the Enlightenment elevated reason to supremacy. The formerly depraved humanity that needed a Savior could, through reason educated by experience, understand the mysteries of God's creation. Mankind became exalted as the conqueror of nature; his potential became unlimited. Paradoxically, the liberation and elevation of mankind also resulted in his demotion from the pinnacle of creation. No longer "a little lower than the angels," man became the product of random chance - on more or less equal status as the worms and bacteria. Eventually this worldview gave birth to the current Age of Modernity.

The epistemological principle of the resulting Modern Age was that human reason, within a context of an objective and rational universe, could master the laws of nature and eventually subdue the universe. This age created a new genre of fiction in which man would "explore strange new worlds, seek out new life and new civilizations, and boldly go where no man has gone before." Such was the confidence and optimism of humanity.

In order to transition from the pre-Enlightenment epistemology of revealed truth to one of experiential truth, philosophers had to develop the concept of natural religion and revealed religion. In Modernity, natural religion was considered acceptable. It contained the fundamentals of ethics written into the mind. These fundamentals could then be accessed through reason. For example, everyone knows (although the Modernist can't prove it) that torturing babies for fun is wrong. What could be wrong with that kind of religion?

Although tolerated in general society, revealed religion was much less acceptable in Modernist society and was banned from most intellectual pursuits. Revealed religion contained all the aspects of religion outside of natural religion, such as the Bible (although it is allowed as a great ethical book and contained a statement of philosophy worthy of study in theology departments). Miracles, such as the creation of the world by the Word of God or the raising of the dead by Jesus, were within the realm of revealed religion, but were not permitted whatsoever within the experiential epistemology.

An outgrowth of the Enlightenment was the scientific revolution. Science has been the most successful method of the Modern Age for gaining truth. It is the goal of science to unlock the secrets of the universe and create a better world. Science has been amazingly successful in doing so as evidenced by our technological society. Therefore, to answer the question, "What is truth?" one should start with an understanding of the question, "What is science?"

What is Science?

Perhaps we all remember what we were taught in elementary school, or even in college, on the scientific method. The trouble is, scientists and philosophers of science are no longer sure what the scientific method is. There are many conflicting answers to the question, "What is science?" Originally, science was what produced certainty whereas non-science produced opinion. But science is not inherently epistemologically superior; it too produces uncertainty. We have been taught that science is based on deductive proofs, but almost always it works by inductive inferences that can be grossly wrong. Science actually operates on the assumption that theories must be falsified so we can move on to better theories - ever approaching truth but never attaining it.

Next, science was distinguished by its unique method. But, in spite of Bacon and our elementary textbooks, there is no rigorous scientific method that can be prescribed. The next criterion thought to distinguish science was Sir Karl Popper's falsifiability. However, even this demarcation between science and non-science has failed. As much as we would like to do so, we cannot even rigorously eliminate astrology from science. In reality, science is what society determines it is. We are generally able to sense what science is and is not, but we cannot rigorously define it. Under these circumstances, academics and other intelligentsia get to subjectively define science; it is what they allow it to be. In that regard, the epistemological principles of science are not much different than those of revelation.

Under the tutelage of scientist-philosopher Michael Ruse, Judge William Overton attempted to establish criteria for what was allowed within the bounds of science. Judge Overton's rules for defining science are contained in his 1982 decision on McLean v. Arkansas Board of Education (Johnson 1993):

1. It is guided by natural law.

2. It has to be explanatory by reference to natural law.

3. It is testable against the empirical world.

4. Its conclusions are tentative - that is, not necessarily the final word.

5. It is falsifiable.

The first two criteria can be addressed together; they are an expression of naturalism. Many laws are descriptive but not explanatory. For example, the law of gravitation states that the gravitational force between two bodies is proportional to the product of their masses and inversely proportional to the square of the distance between them. Although such a law may seem intuitive, scientists do not understand why the law holds true.

Conversely, many scientific explanations do not use laws, especially in the historical sciences. These sciences, such as cosmology, archaeology, evolutionary biology, historical geology, and paleontology, focus on past events that are neither observable nor repeatable. They use present-day observations to extrapolate into history the causes of current conditions. For example, evolution is the postulating of hypothetical historical events that, if true, would explain some of the present-day observations. However, such extrapolations do not rely on natural law. They are historical stories established in the tradition of science fiction to explore hypothetical origins.

Therefore, are historical sciences, such as biological evolution, more scientific than postulates of an intelligent designer? Causation through the agency of some purposeful entity, whether human or divine, does not necessarily place the question of origins outside the bounds of science. For example, does forensic science violate the laws of nature or exist beyond science just because an historical event under consideration was designed and implemented by an intelligent being? Clearly, Overton's rules 1 and 2 cannot distinguish science since so much of science does not conform to the definition.

Overton's criteria three through five are the essence of empiricism. Yet even these defining characteristics are routinely violated by today's science. Within the historical sciences, there are many examples of non-testable hypotheses, including natural selection as a cause for macroevolution. While it is true that scientific theories are supposed to be tentative, always subject to revision, one will not find many scientists willing to state that descent with modification is tentative. ("Evolution is a fact, fact, FACT!" see Chapter 5). Is this as dogmatic or more so than creationism?

Finally, falsifiability is an excellent criterion for a theory when it is attainable. But philosophers of science have come to realize that very few scientific propositions are falsifiable. Scientific predictions (which are infrequently made) are not just based on the core principles of a theory but on a combination of core principles and auxiliary hypotheses. Therefore, when a seemingly falsifying experiment is performed, the scientist is able to devise an *ad hoc* explanation to maintain the theory's standing. That is why Kuhn says that large numbers of anomalies do not falsify a paradigm; the paradigm must be replaced by an entirely new paradigm first (Kuhn 1996).

The central paradigm of science today is naturalism. While science could profitably be pursued without naturalism as its basis, the combination of naturalism and scientific inquiry leads to the practice of scientism. Scientism states that science is the only path to truth and that science ultimately has no limits. However, science itself is founded on philosophy. It must be established on philosophical presuppositions such as (1) the world is an orderly and knowable place, (2) there is such a thing as absolute truth, (3) the laws of logic exist and are reliable, and (4) values are important to the accurate pursuit of science (i.e., no falsifying of data, etc.).

Therefore, how could science be superior to philosophy? Scientism is, thus, self-refuting. Indeed, some philosophical statements outside the bounds of science are more fundamentally correct than many within science. This implies that a search for truth should involve all relevant and appropriate methods including theology, philosophy, and science. As wonderful as science has been for humankind, it is a defective means of acquiring truth when employed in isolation.

Kuhn has shown us that scientific truth is largely a social phenomenon. The reigning paradigm determines what scientists allow themselves to see when they view the world. It governs what questions scientists are willing to ask, which experiments will be funded and performed. The halting, hit and miss progress of science under the misdirection of paradigm-induced misperception cannot be the ideal way to approach truth. Perhaps a better way is to utilize revealed truth, but the big question is, why should we trust revealed truth?

Apologetics

Humans seem to have built into them a longing for something better and a desire for understanding of their origins. The universal existence of religions throughout all times and cultures testifies to this fact[37]. Even atheists have adopted the religion of naturalism. In these religions, there are many views of the Creator and His creation. The view most literally expressed in the Bible is the one held by the majority of Christians, Hebrews, and Muslims - that God is a personal entity, transcendent of time and material existence, and that He has created the universe through a direct and planned action. God has also indirectly maintained the universe by establishing physical laws to govern matter, energy, and time. Many theists also believe that God intervenes in this world in times and manners of His own choosing.

The message of Chapter 4 of this book is that nature contains strong evidence of an intelligent designer. If a personal God created the universe and inhabited it with intelligent beings designed to inquire into their environment and their origins, perhaps creation itself would be

[37] With blatant disregard for the precepts of multiculturalism, one might recognize that not all religions are good - many have been outright evil. Neither the goodness nor the accuracy of religion is important to the point here that humans are inherently religious.

a message from the Creator of His existence. The Bible states this principle that nature and science are a revelation of God's truths:

> *The heavens declare the glory of God; and the firmament sheweth his handywork. Day unto day uttereth speech, and night unto night sheweth knowledge.* Psalms 19:1, 2.

> *When I consider thy heavens, the work of thy fingers, the moon and the stars, which thou hast ordained; What is man that thou are mindful of him? And the son of man, that thou visitest him? For thou hast made him a little lower than the angels, and hast crowned him with glory and honor. Thou madest him to have dominion over the works of thy hands; thou hast put all things under his feet.* Psalms 8:3-6.

> *For the invisible things of him from the creation of the world are clearly seen, being understood by the things that are made, even his eternal power and Godhead; so that they are without excuse.* Romans 1:20.

However, as we briefly discussed in Chapter 7, nature and mankind have been degraded from the original creation. Although important faculties that should be applied to all endeavors, human logic and reasoning in isolation are no longer reliable guides to truth. Nature no longer testifies as clearly to the principles of God as it once did. Therefore, our understanding of the "book of nature" is incomplete. We draw false conclusions about God and his creation from study of science alone.

If the Creator really exists, as the evidence for intelligent design implies, and if He is really a personal being interested in humanity, perhaps He also gave a more direct revelation than nature. The Bible claims to be that revelation.

> *Thy word is a lamp unto my feet, and a light unto my path.* Psalms 119:105.

> *All Scripture is given by inspiration of God, and is profitable for doctrine, for reproof, for correction, for instruction in righteousness.* 2 Timothy 3:16.

Correctly understood, the book of nature and the book of inspiration should complement each other. There should be no contradictions since God, the Creator of the universe, is the author of both. The God who created nature has taken a great interest in humanity by providing Scripture to guide our inquiries into science and the meaning of life. We do not have to be led astray by the imperfections of naturalistic epistemology.

A Holistic Epistemology

Humans can be wrong about our religion, our science, and our philosophy. How then can we know truth? Perhaps at this time in Earth's history, there is no absolute certainty about anything (that is not to say there are no absolutes). However, as revealed truth, science, and philosophy combine in a consistent structure pointing to the same truth, we can have greater confidence in its accuracy. (Nevertheless, even our understanding of the Bible should be tested to see if it is reasonable to believe. In the end, belief is a matter of faith. This discussion in no way attempts to prove the existence of God.)

Scientists who are Christians[38] should consult all they know in their investigation of science. This would include both scientific investigation and revealed truth. But which has supremacy? Revealed truth cannot be used as scientific truth, but it can certainly guide scientific inquiry. Conversely, scientific propositions should not be used to nullify revealed truth. Revealed truth is, after all, <u>direct</u> revelation while nature is more implied revelation. Although the Bible can be misunderstood and misinterpreted, the plain, literal statements that are consistently presented throughout Scripture should hold supremacy over scientific propositions. If there is an apparent conflict, it will be resolved one day by more accurate understanding of science. It is this supremacy granted to revealed truth that infuriates atheistic scientists the most.

Thus, we are suggesting a holistic epistemology in which revealed truth is allowed to guide scientific inquiry. In this construct of scientific inquiry, "miracles" would be allowed. But miracles are nothing more than the intervention of divine agency in the affairs of the universe. They are applications of principles that we do not understand, fundamentally the same as singularities such as the big bang that scientists are accustomed to dealing with. Scientists derisively call this the "God of the gaps" approach to science in which God's miracles fill in what we do not understand, but with the gaps closing to infinitesimal dimensions as science progresses.

---

[38] The reader will note that by selecting the Holy Bible and Christianity as the norm for religion, the author is revealing his bias. The argument for selecting Christianity as superior to the other religions of the world is beyond the scope of this discussion. Interested parties may contact the author for a personal testimony on how he came to know Jesus.

However, holistic epistemology, or theistic science as it is sometimes called, is not God-of-the-gaps thinking. It is merely recognition of the possibility of causation by principles beyond our understanding. Just because science cannot now (or perhaps ever) inquire into supernatural agency, does theistic science necessarily fall outside the bounds of science? To ignore the possibility of supernatural causation would be ludicrous if there really is an intelligent designer (and science absolutely cannot exclude the possibility that there is). The unfalsifiable hypotheses inherent in theistic science are no more unworthy of science than those inherent in naturalism (i.e., the *a priori* assumption that there is no Creator).

How differently would science be conducted today if theistic science were permitted? What can revelation offer science? Currently, nearly all pure research NASA does relates to finding life where none should exist and to understanding the cosmology of the big bang that occurred much differently than cosmologists currently suppose. The SETI program works to find messages in the stars, when there are messages in revealed truth that are being ignored. It is difficult to imagine how much funding and effort is directed towards research programs that cannot bear fruit, according to revelation. This huge effort is based on an *a priori* assumption that the universe made itself. Theistic science could redirect research into more fruitful areas. Perhaps the fingerprints of God's singularities (miracles) could be found in nature if scientists were to look for them.

Appeal

The Scriptures are very blunt about the philosophy of naturalism, "*The fool hath said in his heart, There is no God.*" Psalm 53:1. In these eight chapters, we have attempted to show that the atheistic approach to science, called naturalism, is not a requirement of science but an unfounded presupposition, that nature reveals the actions of an intelligent designer, that we do not need to become fools.

Intellectual history indicates that mankind has step-by-step built the case for naturalism, almost purposefully eliminating the Creator from considerations of reality. Given the pressure to accept naturalism, it is a wonder that any intellectuals go against the prevailing opinion and accept divine origin for the universe. However, every individual has the opportunity to confront the evidence and draw his or her own conclusion. The Bible reveals that God wants to reach every human on the Earth with his message of salvation, to redeem humanity from the degradation resulting from transgression of His Law. The author would urge every reader to seek the Creator's message in nature and revelation.

> *For I know the thoughts that I think toward you, saith the Lord, thoughts of peace, and not of evil ... And ye shall seek me, and find me, when ye shall search for me with all your heart.* Jeremiah 29:11, 13.

Nine

References

Chapter 1

Darwin, Charles R., 1859. *The Origin of Species*, 1st Edition, reprint by Bantam Books in 1999, New York.

Dawkins, Richard, 1996. *The Blind Watchmaker: Why the Evidence of Evolution Reveals a Universe Without Design*, W. W. Norton and Company, London.

Denton, Michael, 1985. *Evolution: A Theory in Crisis*, Adler & Adler, Bethesda, Maryland.

Britannica, 1999. "Evolution," article in Encyclopedia Britannica as found at www.britannica.com, September 1999.

Johnson, Phillip E., 1993. *Darwin on Trial*, Intervarsity Press, Downers Grove, Illinois.

www.greats2000.org/parsed/data/articles/HarveyWilliam.html

www.lucknow.com/horus/student/harvey.html

Chapter 2

Behe, Michael J., 1996. *Darwin's Black Box*, Touchstone, New York, New York.

Darwin, Charles R., 1859. *The Origin of Species*, 1st Edition, reprint by Bantam Books in 1999, New York, New York.

Denton, Michael, 1985. *Evolution: A Theory in Crisis*, Adler & Adler, Bethesda, Maryland.

Encyclopedia Britannica, 2000. as found at www.britannica.com.

Futuyma, Douglas J., 1998. *Evolutionary Biology*, 3rd Edition, Sinauer Associates, Inc., Sunderland, Massachusetts.

Johnson, Phillip E., 1993. *Darwin on Trial*, Intervarsity Press, Downers Grove, Illinois.

Chapter 3

Behe, Michael J., 1996. *Darwin's Black Box, the Biochemical Challenge to Evolution*, Touchstone, New York, New York.

Britannica, 2000. *Encyclopedia Britannica*, article, "Life," as found at www.britannica.com in April 2000.

Darwin, F., 1887. *The Life and Letters of Charles Darwin*, Appleton, New York, New York.

Denton, Michael, 1985. *Evolution: A Theory in Crisis*, Adler & Adler, Publishers, Bethesda, Maryland.

Moreland, J. P., editor, 1994. *The Creation Hypothesis, Scientific Evidence for an Intelligent Designer*, Intervarsity Press, Downers Grove, Illinois.

Chapter 4

Beye, Michael J., 1996. *Darwin's Black Box*, Touchstone, New York, New York.

Connor, S. J., undated. "Why Bad Things Happen to Good People," unpublished essay, available on request, Aiken, South Carolina.

Dawkins, Richard, 1996. *The Blind Watchmaker: Why the Evidence of Evolution Reveals a Universe Without Design*, W. W. Norton and Company, London.

Denton, Michael, 1985. *Evolution: A Theory in Crisis*, Adler & Adler, Bethesda, Maryland.

Denton, Michael, 1998. *Nature's Destiny - How the Laws of Biology Reveal Purpose in the Universe*, The Free Press, New York, New York.

Futuyma, Douglas J., 1998. *Evolutionary Biology*, 3rd Edition, Sinauer Associates, Inc., Sunderland, Massachusetts.

Johnson, Phillip E., 1993. *Darwin on Trial*, Intervarsity Press, Downers Grove, Illinois.

Kuhn, Thomas S., 1996. *The Structure of Scientific Revolutions*, 3rd edition, the University of Chicago Press, Chicago, Illinois.

Nelson, Paul A., 1998. "Applying Design Within Biology," in *Mere Creation - Science, Faith & Intelligent Design*, edited by William A. Dembski, Intervarsity Press, Downers Grove, Illinois.

Numbers, Ronald L., 1993. *The Creationists - The Evolution of Scientific Creationism*, University of California Press, Los Angeles, California.

Regis, Ed, 1995. *Nano - The Emerging Science of Nanotechnology*, Little, Brown, and Company, Boston, Massachusetts.

Chapter 5

Anderson, Norris, 1995. *Education or Indoctrination - Analysis of Textbooks in Alabama*, Norris Anderson, reprinted at www.arn.org/docs/anderson/analysis_main.htm.

Britannica 2000. *Encyclopedia Britannica*, article "Plate Tectonics," www.britannica.com, accessed June 2000.

Denton, Michael, 1985. *Evolution: A Theory in Crisis*, Adler & Adler, Bethesda, Maryland.

Futuyma, Douglas J., 1998. *Evolutionary Biology*, 3rd Edition, Sinauer Associates, Inc., Sunderland, Massachusetts.

Gentry, Robert V., 1986. *Creation's Tiny Mystery*, Earth Science Associates, Knoxville, Tennessee.

Johnson, Phillip E., 1993. *Darwin on Trial*, Intervarsity Press, Downers Grove, Illinois.

Johnson, Phillip E., 1998. *Objections Sustained - Subversive Essays on Evolution, Law & Culture*, Intervarsity Press, Downers Grove, Illinois.

Kuhn, Thomas S, 1996. *The Structure of Scientific Revolutions*, 3rd edition, The University of Chicago Press, Chicago, Illinois.

Popper, Karl R., 1989. *Conjectures and Refutations - the Growth of Scientific Knowledge*, Routledge, London, England, reprinted in 1996.

Ruse, Michael, 1982. *Darwinism Defended - A Guide to the Evolution Controversies*, Addison Wesley, Reading, Pennsylvania.

Taylor, Ian T., 1991. *In the Minds of Men - Darwin and the New World Order*, 3rd edition, TFE Publishing, Minneapolis, Minnesota.

Chapter 6

Austin, Steven A., 1993. *Mt. St. Helens - Explosive Evidence for Catastrophe*, video tape, Institute for Creation Research, El Cajon, California.

Austin, Steven A., editor, 1994. *Grand Canyon - Monument to Catastrophe*, Institute for Creation Research, Santee, California.

Austin, S. A., 1996. "Excess Argon within Mineral Concentrates from the New Dacite Lava Dome at Mount St. Helens Volcano," *Creation Ex Nihilo Technical Journal*, Vol 10, No. 3, page 355.

Barber, Mike, 2000. "Crotchety Harry Truman Remains an Icon of the Eruption," Seattle Post-Intelligencer, Seattle, Washington, May 11.

Barnes, Virgil E. and Mildred A. Barnes, editors, 1973. *Tektites*, Dowden, Hutchinson, & Ross, Inc., Stroudsburg, Pennsylvania.

Brown, Walt, 1995. *In the Beginning: Compelling Evidence for Creation and the Flood*, 6th edition, Center for Scientific Creation, Phoenix, Arizona.

Gentry, Robert V., 1992. *Creation's Tiny Mystery*, 3rd edition, Earth Science Associates, Knoxville, Tennessee.

Sewell, Curt, 1997. *God at Ground Zero*, Master Books, Inc., Green Forest, Arkansas.

Taylor, Ian T., 1991. *In the Minds of Men - Darwin and the New World Order*, Third Edition, TFE Publishing, Minneapolis, Minnesota.

Chapter 7

CNN (Cable News Network) 2000. "Fossilized Dinosaur Heart Rocks Paleontology," www.cnn.com/2000/TECH/computing/05/11/dino.heart.t t/index.html, May 11, 2000.

Darwin, Charles R., 1859. *The Origin of Species*, 1st Edition, reprint by Bantam Books in 1999, New York.

Encyclopedia Britannica, articles on human evolution, hominid, dinosaur, and horse, as found at www.britannica.com, August 2000.

Futuyma, Douglas J., 1998. *Evolutionary Biology*, 3rd Edition, Sinauer Associates, Inc., Sunderland, Massachusetts.

Taylor, Ian T., 1991. *In the Minds of Men - Darwin and the New World Order*, Third Edition, TFE Publishing, Minneapolis, Minnesota.

Wieland, Carl 1998. "Dinosaur Bones: Just How Old are They Really?", *Creation Ex Nihilo* **21**(1) 54-55, December 1998-February 1999, Answers in Genesis Ministries, Florence, Kentucky.

Chapter 8

Grenz, Stanley J, 1996. *A Primer on Postmodernism*, Wm. B. Eerdmans Publishing Co., Grand Rapids, Michigan.

Johnson, Phillip E., 1993. *Darwin on Trial*, Intervarsity Press, Downers Grove, Illinois.

Kuhn, Thomas S., 1996. *The Structure of Scientific Revolutions*, 3rd edition, The University of Chicago Press, Chicago, Illinois.

Moreland, J. P., editor, 1994. *The Creation Hypothesis, Scientific Evidence for an Intelligent Designer*, Intervarsity Press, Downers Grove, Illinois.